Mathematics for Industry

Volume 34

Aims & Scope

The meaning of "Mathematics for Industry" (sometimes abbreviated as MI or MfI) is different from that of "Mathematics in Industry" (or of "Industrial Mathematics"). The latter is restrictive: it tends to be identified with the actual mathematics that specifically arises in the daily management and operation of manufacturing. The former, however, denotes a new research field in mathematics that may serve as a foundation for creating future technologies. This concept was born from the integration and reorganization of pure and applied mathematics in the present day into a fluid and versatile form capable of stimulating awareness of the importance of mathematics in industry, as well as responding to the needs of industrial technologies. The history of this integration and reorganization indicates that this basic idea will someday find increasing utility. Mathematics can be a key technology in modern society.

The series aims to promote this trend by 1) providing comprehensive content on applications of mathematics, especially to industry technologies via various types of scientific research, 2) introducing basic, useful, necessary and crucial knowledge for several applications through concrete subjects, and 3) introducing new research results and developments for applications of mathematics in the real world. These points may provide the basis for opening a new mathematics-oriented technological world and even new research fields of mathematics.

To submit a proposal or request further information, please use the PDF Proposal Form or contact directly: Swati Meherishi, Executive Editor (swati.meherishi@springer.com).

More information about this series at http://www.springer.com/series/13254

Hiromichi Itou · Shiro Hirano ·
Masato Kimura · Victor A. Kovtunenko ·
Alexandr M. Khludnev
Editors

Mathematical Analysis of Continuum Mechanics and Industrial Applications III

Proceedings of the International Conference CoMFoS18

Springer

Editors
Hiromichi Itou
Department of Mathematics
Tokyo University of Science
Tokyo, Japan

Shiro Hirano
College of Science and Engineering
Ritsumeikan University
Kusatsu, Japan

Masato Kimura
Faculty of Mathematics and Physics
Kanazawa University
Kanazawa, Ishikawa, Japan

Victor A. Kovtunenko
NAWI Graz, Institut für Mathematik
Karl-Franzens University of Graz
Graz, Austria

Alexandr M. Khludnev
Siberian Branch of RAS
Lavrentyev Institute of Hydrodynamics
Novosibirsk, Russia

ISSN 2198-350X ISSN 2198-3518 (electronic)
Mathematics for Industry
ISBN 978-981-15-6061-3 ISBN 978-981-15-6062-0 (eBook)
https://doi.org/10.1007/978-981-15-6062-0

This Springer imprint is published by the registered company Springer Nature Singapore Pte Ltd.
The registered company address is: 152 Beach Road, #21-01/04 Gateway East, Singapore 189721, Singapore

Preface (CoMFoS18)

Continuum Mechanics Focusing on Singularities (CoMFoS) is the name of a conference series currently chaired by founding member Kohji Ohtsuka from the Hiroshima Kokusai Gakuin University. The CoMFoS conference has continued for more than 23 years as an activity of Mathematical Aspects of Continuum Mechanics (MAMC), a research group that is part of the Japan Society for Industrial and Applied Mathematics.

In 2018, the 18th international conference (CoMFoS18) on 'Mathematical Analysis of Continuum Mechanics II' was held at Shirankaikan Annex, Kyoto, Japan, from June 13 to 15. It was supported by the Inoue Foundation for Science and the Japan Society for the Promotion of Science (JSPS) KAKENHI. This conference was co-organized by the Research Institute for Science and Technology at the Tokyo University of Science, 'Division of Mathematical Modelling and its Mathematical Analysis', Kanazawa Mathematics and mathematical science Research Group in Kanazawa University, and the Austrian Science Fund (FWF) research project 'Object identification problems: Numerical analysis (PION)'. We are very grateful to all supports; without their assistance, the conference could never have been such a great success.

In recent years, remarkable progress has been made in research because of the development of sensing technologies and computer resources. However, with the evolution of technologies, most studies have tended to become specialized and subdivided. To understand the various phenomena occurring in our surroundings, it is necessary to examine each phenomenon from a higher perspective and theoretically integrate the diverse developments in the various research fields. Mathematics plays a crucial role in integrating diverse scientific fields and industries as the common language. For CoMFoS18, we used shared concepts to bring researchers from all over the world on a common platform; the researchers were from diverse fields such as mathematics, engineering and seismology. This conference focussed on topics that continue to be the most important concern in Japan, such as earthquakes. Earthquakes are complicated and composite phenomena. During an earthquake, fault destruction occurs after the deformation of the plate considered as elasticity or viscoelasticity. Subsequently, seismic waves and tsunamis are generated and propagated, which cause some structures to vibrate and be destroyed. Therefore, appropriate strategies are required for not only fracture problems, but also shape optimization, inverse problems. The CoMFoS18 conference provided a platform for researchers to discuss the latest scientific results; it also helped improve the interactions among researchers from various fields.

We are very pleased to publish the proceeding of CoMFoS18 as part of the 'Mathematics for Industry' series and as a sequel to the proceedings of CoMFoS15 and CoMFoS16. Further, we would like to express our gratitude to the Institute of Mathematics for Industry of Kyushu University, the editorial staff at Springer publication and JSPS and Russian Foundation for Basic Research (RFBR) research projects J19-721 and 16-51-50004 for support. This book contains 14 selected papers that bring together a wide variety of viewpoints. It includes a survey of the basic concepts behind various phenomena, mathematical modelling, computer simulation methodology, theoretical results, and so on. We hope that this book will inspire many scientists, upcoming researchers, in particular, to take up the challenge of investigating hitherto unexplained phenomena.

Tokyo, Japan

Hiromichi Itou
On behalf of the Organizing
Committee of CoMFoS18

Contents

Fracture Mechanics and Radiative Transport

Dynamic Unilateral Contact Problem with Averaged Friction for a Viscoelastic Body with Cracks

Atusi Tani

Abstract In order to study the mechanism of earthquake through fracture mechanics, as a first step we discuss a dynamic unilateral contact problem with friction for a cracked viscoelastic body. Here we adopt the viscoelastic model proposed by Landau and Lifshitz [24] and a non-local friction law. The existence of a solution to the problem is obtained by using a penalty method. Several estimates are obtained on the solution to the penalized problem, which enable us to pass to the limit by using compactness results [30].

1 Introduction

Roughly speaking, earthquakes occur as a result of global plate motion [21]; an earthquake is a sudden rupture process in the Earth's crust or mantle caused by tectonic stress. To understand the physics of earthquakes it is important to determine the state of stress before, during, and after an earthquake [20]; earthquakes may be considered to be dynamically running shear cracks [29]. In general, all real materials contain defects. Two types of defects are important: cracks, which are surface defects, and dislocations, which are line defects. These two mechanisms result in grossly different macroscopic behaviors. When cracks are the active defect, material failure occurs by its separation into parts, often catastrophically: this is brittle behavior. Plastic flow results from dislocation propagation, which produces permanent deformation without destruction of the lattice integrity.

In this paper, we study the dynamic crack motion via a fracture mechanics (or a continuum mechanics) approach, in which the crack is idealized as a mathematically flat and narrow slit in a linear viscoelastic medium. It consists of analyzing the stress field around the crack, and then formulating a fracture criterion based on certain critical parameters of the stress field. The macroscopic strength is thus related to the intrinsic strength of the material through the relationship between the applied stresses

A. Tani (✉)
Department of Mathematics, Keio University, 3-14-1 Hiyoshi, Yokohama 223-8522, Japan
e-mail: tani@math.keio.ac.jp

H. Itou et al. (eds.), *Mathematical Analysis of Continuum Mechanics and Industrial Applications III*, Mathematics for Industry 34,
https://doi.org/10.1007/978-981-15-6062-0_1

and the crack tip stresses. Because the crack is treated as residing in a continuum, the details of the deformation and fracturing processes at the crack tip are ignored (see [3, 15]). Shear failure under compressive stress states is commonly described with the "Coulomb criterion" (or Navier–Coulomb or Coulomb–Mohr criterion), which evolves from the simple frictional criterion for the strength of cohesionless soils. Friction is the resistance to motion that occurs when a body slides tangentially to surface on which it contacts another body. It plays an important role in a great variety of processes. It is always present in mechanics in which there are moving parts (see [18, 27]).

There are a few mathematical results dealing with the existence of a solution to modified dynamic unilateral contact problems with friction for viscoelastic bodies, *e.g.*, [5–10, 13, 19, 26, 28]. However, all of them are far from the exact solvability results for the dynamic unilateral contact problems with friction for not only viscoelastic but also elastic bodies. When we consider these problems in the weak framework, the stress tensor does not have sufficient regularity on the boundary for the contact boundary conditions to make sense. To overcome this difficulty in [13], authors discussed such a problem under the contact condition in velocities, not in displacements, and other referred authors employed Coulomb friction law with non-local friction (starting from [11]) or normal compliance. The non-local friction model compactifies the friction terms by mollifying the normal force in it, while the tangential ones remain mostly unchanged. Certainly, no physical base for both treatments has been presented until now. Concerning the latter models, they involve one or several small parameters which can be physically justified with the help of a micro-scale analysis. The micro-scale analysis fixes the size of the small parameter to be of the order of magnitude of the roughness of the boundary, which leads to the physically reasonable non-local Coulomb friction law or the improved normal compliance model. Surely, a formal passage to the limit in the normal compliance model gives a contact problem with Signorini contact condition and Coulomb friction law, so that the important qualitative questions about the normal compliance models are essentially equivalent to the analysis of Coulomb friction: if the solution of a normal compliance model does not depend on the small parameter involved in this model, i.e., some kind of penalty approximation to Coulomb friction converges in practice.

Nevertheless, here we adopt Coulomb friction law with non-local friction. The aim of this paper is to present a result on the entitled topic, which is found much similar to that in [9] after completed: as a model of viscoelastic bodies we adopt the one due to Landau and Lifshitz [24] in this paper, while in [9] the one of Kelvin–Voigt type was used. Once one follows the arguments thanks to Duvaut and Lions in [11, 12], it seems to be like one another, as an inevitable consequence, in proving the existence results.

2 Formulation of the Problem and Main Theorem

Let Ω be a bounded domain in $\mathbb{R}^3$. We regard a viscoelastic body as an isotropic elastic body with viscosity proposed by Landau and Lifshitz [24] which initially occupies Ω. Suppose that the boundary $\partial\Omega$ of Ω is composed of three parts, $\partial\Omega = \bar{\Gamma}_U \cup \bar{\Gamma}_F \cup \bar{\Gamma}$, where Γ_U and Γ_F are sufficiently smooth and $\text{meas}(\Gamma_U) > 0$; the crack $\Gamma = \Gamma^+ \cup \Gamma^-$, which may be in unilateral contact with friction.

The motion of such a body is described by

$$\rho\,\frac{\partial \mathbf{v}}{\partial t} = \nabla\cdot\mathscr{P} + \rho\mathbf{f}, \quad \mathbf{v} = \frac{\partial \mathbf{u}}{\partial t} \quad \text{in } \Omega,\ t > 0. \tag{1}$$

Here $\mathbf{v}$ is a velocity vector field, $\mathbf{u}$ is a displacement vector field, $\mathbf{f}$ is a body force, ρ is the density (constant) of the body, $\mathscr{P}$ is a stress tensor defined by

$$\begin{aligned}
&\mathscr{P} = \mathscr{P}(\mathbf{u},\mathbf{v}) = \mathscr{P}^{(\mathrm{el})} + \mathscr{P}^{(\mathrm{vis})},\\
&\mathscr{P}^{(\mathrm{vis})} = \mathscr{P}^{(\mathrm{vis})}(\mathbf{v}) = 2\mu\,\mathscr{D}(\mathbf{v}) + \mu'\,\mathrm{tr}(\mathscr{D}(\mathbf{v}))\,\mathscr{I},\\
&\mathscr{P}^{(\mathrm{el})} = \mathscr{P}^{(\mathrm{el})}(\mathbf{u}) = 2\lambda\,\mathscr{D}(\mathbf{u}) + \lambda'\,\mathrm{tr}(\mathscr{D}(\mathbf{u}))\,\mathscr{I}
\end{aligned}$$

with $\mu,\ \mu'$, the fluid viscosities and $\lambda,\ \lambda'$, Lamé constants satisfying $\mu > 0,\ 3\mu' + 2\mu \geq 0,\ \lambda > 0,\ 3\lambda' + 2\lambda \geq 0,\ \mathscr{D}(\mathbf{w}) = (\nabla\mathbf{w} + \nabla\mathbf{w}^{\mathrm{T}})/2$, the deformation tensor, and $\mathscr{I}$, an identity tensor (see [24, 25]).

First, we introduce the parametrization of the crack by the following [1].

Let $\Omega = \Omega^+ \cup \Omega^- \cup \Gamma_V$, $\Omega^+ \cap \Omega^- = \emptyset$, with Lipschitz continuous boundaries $\partial\Omega^\pm$, and $\Gamma_V \subset \partial\Omega^+ \cap \partial\Omega^-$ may be regarded as a virtual interface between Ω^+ and Ω^-. We choose the above decomposition such that $\text{meas}(\Gamma_U^\pm) > 0$, where $\Gamma_U^\pm = \Gamma_U \cap \partial\Omega^\pm$, and we denote $\Gamma^\pm = \Gamma \cap \partial\Omega^\pm$.

Now we introduce a reference domain Ξ in $\mathbb{R}^2$, independent of Ω^+ and Ω^-, to express boundary conditions on the crack. Assume that $\Gamma^\pm$ are defined as $\Gamma^\pm = \left\{\left(\bar{x}, \phi^\pm(\bar{x})\right) \mid \bar{x} \in \Xi\right\}$ by $\phi^\pm \in C^1(\Xi)$ such that $\phi^+(\bar{x}) - \phi^-(\bar{x}) \geq 0$ for any $\bar{x} \in \Xi$.

Let $\mathbf{y}(t,\mathbf{x})$ be the position at time t of the material point which was initially at $\mathbf{x}$, and $\mathbf{u}(t,\mathbf{x}) = \mathbf{y}(t,\mathbf{x}) - \mathbf{x}$ be the displacement vector field of $\mathbf{x}$ at time t. By $\mathbf{m}^+ = (\nabla\phi^+(\xi), -1)$ and $\mathbf{m}^- = (-\nabla\phi^-(\xi), 1)$ we denote the outward normal to Γ^+ and Γ^-, respectively.

Assume that the contact surface at time t is represented implicitly by $h(t,\mathbf{y}) = 0$ with some function h satisfying $\partial h/\partial y_3 > 0$, and the non-penetrating condition for the crack faces is

$$h(t,\mathbf{y}(t,\mathbf{x}^+)) \geq 0 \quad (\forall\mathbf{x}^+ \in \Gamma^+), \quad h(t,\mathbf{y}(t,\mathbf{x}^-)) \leq 0 \quad (\forall\mathbf{x}^- \in \Gamma^-). \tag{2}$$

Implicit function theorem implies that there exists a function $\psi : (0,T) \times \mathbb{R}^2 \to \mathbb{R}$ such that

$$h(t,\mathbf{y}) = 0 \iff y_3 = \psi(t,\bar{y}), \quad \bar{y} = (y_1, y_2) \in \mathbb{R}^2.$$

This yields that (2) is equivalent to $y_3^+ - \psi(t, \bar{y}^+) \geq 0$, $y_3^- - \psi(t, \bar{y}^-) \leq 0$, where $\bar{y}^\pm = (y_1(t, \mathbf{x}^\pm)), y_2(t, \mathbf{x}^\pm)) = \bar{x}^\pm + \bar{u}^\pm$, $\bar{x}^\pm = (x_1^\pm, x_2^\pm)$, $\bar{u}^\pm = (u_1(t, \mathbf{x}^\pm)), u_2(t, \mathbf{x}^\pm))$. Then, we obtain, by setting $\bar{x}^+ = \bar{x}^- = \xi$,

$$x_3^+ + u_3^+(t, \mathbf{x}^+) - \psi(t, \xi + \bar{u}^+) \geq 0, \quad x_3^- + u_3^-(t, \mathbf{x}^-) - \psi(t, \xi + \bar{u}^-) \leq 0. \quad (3)$$

Throughout this paper, the small deformation hypothesis is assumed to be valid. Thus, keeping only linear terms in (3), we obtain

$$\begin{aligned} x_3^+ + u_3^+(t, \mathbf{x}^+) - \psi(t, \xi) - \nabla\psi(t, \xi) \cdot \bar{u}^+ \geq 0, \\ x_3^- + u_3^-(t, \mathbf{x}^-) - \psi(t, \xi) - \nabla\psi(t, \xi) \cdot \bar{u}^- \leq 0, \end{aligned}$$

from which it follows

$$\begin{aligned} &\phi^+(\xi) - \phi^-(\xi) + u_3^+(t, \xi, \phi^+(\xi)) - u_3^-(t, \xi, \phi^-(\xi)) \\ &\quad -\nabla\psi(t, \xi) \cdot \bar{u}^+(t, \xi, \phi^+(\xi)) + \nabla\psi(t, \xi) \cdot \bar{u}^-(t, \xi, \phi^-(\xi)) \geq 0. \end{aligned} \quad (4)$$

Since the gradients can be assumed to be approximately the same,

$$\nabla\phi^+(\xi) \simeq \nabla\psi(t, \xi) \simeq \nabla\phi^-(\xi). \quad (5)$$

[N.B.: It may hold when the gap between the two faces is small.]

Using (4) and (5), we get

$$\begin{aligned} &\phi^+(\xi) - \phi^-(\xi) + u_3^+(t, \xi, \phi^+(\xi)) - u_3^-(t, \xi, \phi^-(\xi)) \\ &\quad -\nabla\phi^+(\xi) \cdot \bar{u}^+(t, \xi, \phi^+(\xi)) + \nabla\phi^-(\xi) \cdot \bar{u}^-(t, \xi, \phi^-(\xi)) \geq 0 \end{aligned}$$

for any $\xi \in \Xi$, or by setting $\mathbf{u}^\pm = (\bar{u}^\pm, u_3^\pm)$,

$$\mathbf{m}^+(\xi) \cdot \mathbf{u}^+(t, \xi, \phi^+(\xi)) + \mathbf{m}^-(\xi) \cdot \mathbf{u}^-(t, \xi, \phi^-(\xi)) \leq \phi^+(\xi) - \phi^-(\xi). \quad (6)$$

This is the unilateral contact condition on Ξ. $\phi^+(\xi) - \phi^-(\xi)$ means the gap between the two faces. By the unit outward normal $\mathbf{n}^\pm = \mathbf{m}^\pm/|\mathbf{m}^\pm|$ to $\Gamma^\pm$, (6) is written as

$$[u_N](t, \xi) \leq g(\xi), \quad \forall\xi \in \Xi, \quad (7)$$

where

$$[w_N](t, \xi) = \mathbf{n}^+(\xi) \cdot \mathbf{w}^+(t, \xi, \phi^+(\xi)) + \mathbf{n}^-(\xi) \cdot \mathbf{w}^-(t, \xi, \phi^-(\xi)), \quad \forall \mathbf{w} = \mathbf{w}(t, \mathbf{x}),$$

$$g(\xi) = \frac{\phi^+(\xi) - \phi^-(\xi)}{\sqrt{1 + |\nabla\phi^+|^2}} \geq 0, \quad \forall\xi \in \Xi.$$

[N.B.: $g \geq 0$ is the normalized gap between the two crack faces; $g = 0$ corresponds to a cut.]

For any $\xi \in \Xi$ we use the following notation:

$$\begin{cases} \mathbf{u}^{\pm} = \mathbf{u}^{\pm}(t,\xi) = \mathbf{u}(t,\xi,\phi^{\pm}(\xi)), \\ u_N^{\pm} = u_N^{\pm}(t,\xi) = \mathbf{u}^{\pm}(t,\xi)\cdot\mathbf{n}^{\pm}(\xi), \quad \mathbf{u}_\tau^{\pm} = \mathbf{u}_\tau^{\pm}(t,\xi) = \mathbf{u}^{\pm} - u_N^{\pm}\mathbf{n}^{\pm}, \\ [u_N] = [u_N](t,\xi) = u_N^{+} + u_N^{-}, \quad [\mathbf{u}_\tau] = [\mathbf{u}_\tau](t,\xi) = \mathbf{u}_\tau^{+} - \mathbf{u}_\tau^{-}, \\ (\mathscr{P})_N^{\pm} = (\mathscr{P})_N^{\pm}(t,\xi) = \mathscr{P}^{\pm}\mathbf{n}^{\pm}\cdot\mathbf{n}^{\pm}, \\ \mathscr{P}_\tau^{\pm} = \mathscr{P}_\tau^{\pm}(t,\xi) = \mathscr{P}^{\pm}\mathbf{n}^{\pm} - (\mathscr{P})_N^{\pm}\mathbf{n}^{\pm}. \end{cases} \tag{8}$$

A classical unilateral contact problem with averaged friction is formulated as follows:

Problem P_0 : Find $u = u(t,\mathbf{x})$ such that $\mathbf{u}(0) = \mathbf{u}_0$, $\mathbf{v}(0) = \mathbf{v}_0$ on Ω and

$$\begin{cases} \rho\dfrac{\partial \mathbf{v}}{\partial t} = \nabla\cdot\mathscr{P} + \rho\mathbf{f}, \quad \mathbf{v} = \dfrac{\partial \mathbf{u}}{\partial t}, \quad \mathscr{P} = \mathscr{P}^{(\mathrm{vis})} + \mathscr{P}^{(\mathrm{el})} \ \text{ in } (0,T)\times\Omega, \\ \mathbf{u} = 0 \ \text{ on } (0,T)\times\Gamma_U, \quad \mathscr{P}\mathbf{n} = \mathbf{F} \ \text{ on } (0,T)\times\Gamma_F, \\ [u_N] \leq g, \quad (\mathscr{P})_N^{+} = (\mathscr{P})_N^{-} \leq 0, \quad (\mathscr{P})_N^{-}([u_N] - g) = 0 \ \text{ on } (0,T)\times\Xi, \\ \mathscr{P}_\tau^{+} = -\mathscr{P}_\tau^{-}, \quad \left|\mathscr{P}_\tau^{+}\right| \leq f\left|(\mathscr{R}\mathscr{P})_N^{+}\right| \ \text{ on } (0,T)\times\Xi, \\ \quad [\mathbf{v}_\tau] = 0 \ \text{ when } \ \left|\mathscr{P}_\tau^{+}\right| < f\left|(\mathscr{R}\mathscr{P})_N^{+}\right|, \\ \quad \exists\lambda \geq 0 \ \text{ such that } [\mathbf{v}_\tau] = -\lambda\mathscr{P}_\tau^{+} \ \text{ when } \ \left|\mathscr{P}_\tau^{+}\right| = f\left|(\mathscr{R}\mathscr{P})_N^{+}\right|. \end{cases} \tag{9}$$

Here the traction $\mathbf{F}$ is prescribed on $(0,T)\times\Gamma_F$, f is the friction coefficient, and $\mathscr{R}$ is an averaging operator defined later which was introduced first by Duvaut [11] (see also [22, 23]).

The conditions on the crack mean non-penetration of the two crack faces and the unilateral contact conditions (Signorini's conditions). The last relations mean the averaged Coulomb friction law.

Before describing the main theorem let us introduce function spaces:

$\mathscr{H} = L_2(\Omega)$ with a scalar product $(\cdot,\cdot)$ and a norm $|\cdot| := \sqrt{(\cdot,\cdot)}$;

$\mathscr{V} = \left\{\mathbf{v}\in H^1(\Omega)\,|\,\mathbf{v} = 0 \text{ a.e. on } \Gamma_U\right\}$ with a scalar product $\langle\cdot,\cdot\rangle$ and a norm $\|\cdot\| := \sqrt{\langle\cdot,\cdot\rangle}$;

$\mathscr{K} = \left\{\mathbf{v}\in\mathscr{V}\ \middle|\ [v_N] \leq g \text{ a.e. on } \Xi\right\}$.

Assume that

$$\mathbf{u}_0\in\mathscr{K}, \quad \mathbf{v}_0\in\mathscr{V}, \quad \mathbf{F}\in W_\infty^1(0,T;L_2(\Gamma_F)), \quad \mathbf{f}\in W_\infty^1(0,T;\mathscr{H}),$$
$$g\in H_{00}^{1/2}(\Xi), \quad g\geq 0 \text{ a.e. on } \Xi, \quad f\in L_\infty(\Xi), \quad f\geq 0 \text{ a.e. on } \Xi.$$

Let a and b be two bilinear, continuous and symmetric forms defined by

$$\begin{cases} a(\mathbf{v}, \mathbf{w}) = \int_{\Omega} \big(2\mu \mathscr{D}(\mathbf{v}) : \mathscr{D}(\mathbf{w}) + \mu' \operatorname{tr}(\mathscr{D}(\mathbf{v})) \operatorname{tr}(\mathscr{D}(\mathbf{w}))\big)\, dx, \\ b(\mathbf{u}, \mathbf{w}) = \int_{\Omega} \big(2\lambda \mathscr{D}(\mathbf{u}) : \mathscr{D}(\mathbf{w}) + \lambda' \operatorname{tr}(\mathscr{D}(\mathbf{u})) \operatorname{tr}(\mathscr{D}(\mathbf{w}))\big)\, dx, \\ \qquad\qquad\qquad\qquad\qquad\qquad \forall \mathbf{u}, \mathbf{v}, \mathbf{w} \in \mathscr{V}. \end{cases} \tag{10}$$

Using the above hypotheses, we define $\mathbf{L} \in W^1_\infty(0, T; \mathscr{V})$ by

$$\langle \mathbf{L}, \mathbf{w} \rangle = \int_{\Omega} \mathbf{f} \cdot \mathbf{w}\, dx + \int_{\Gamma_F} \mathbf{F} \cdot \mathbf{w}\, d\Gamma, \quad \forall \mathbf{w} \in \mathscr{V},\ \forall t \in [0, T]. \tag{11}$$

The averaging operator $\mathscr{R}$ of $\mathscr{P}$, $\mathscr{R} : [L_{2,\mathrm{sym}}(\Omega)]^{3\times 3} \to [H^1(\Omega)]^{3\times 3}$, is linear and continuous, and satisfies $((\mathscr{R}\mathscr{P})(\mathbf{u}_0, \mathbf{v}_0))^+_N = 0$ and the inequality

$$\|(\mathscr{R}\mathscr{P})(\mathbf{u}, \mathbf{v})\| \le C(|\mathbf{u}| + |\mathbf{v}|), \quad \forall \mathbf{u},\ \mathbf{v} \in \mathscr{V} \tag{12}$$

with some positive constant C. Such an averaging operator is constructed by a number of ways. For example, for $\mathbf{u},\ \mathbf{v} \in H^1(\Omega)$ let $E\mathbf{u},\ E\mathbf{v}$ be the extended functions to $\mathbb{R}^3$ such that they satisfy $\|E\mathbf{u}\|_{H^1(\mathbb{R}^3)} \le C\|\mathbf{u}\|_{H^1(\Omega)}$, $\|E\mathbf{v}\|_{H^1(\mathbb{R}^3)} \le C\|\mathbf{v}\|_{H^1(\Omega)}$ with a positive constant C which is independent of $\mathbf{u}$ and $\mathbf{v}$, and define

$$(\mathscr{R}\mathscr{P})(\mathbf{u}, \mathbf{v}) \equiv \mathscr{P}(E\mathbf{u} * \omega, E\mathbf{v} * \omega),$$

where ω is a smooth function with compact support, and $*$ denotes the convolution operation. Thus, $(\mathscr{R}\mathscr{P}) \in C^\infty(\mathbb{R}^3)$ and $(\mathscr{R}\mathscr{P})_N \equiv (\mathscr{R}\mathscr{P})\mathbf{n} \cdot \mathbf{n}$ is well defined on Γ. In this way we average the displacement and the velocity vectors, and consider the stress determined by the averaged variables. Then, the operator $\mathscr{R}$ is linear and continuous, and has the abovementioned properties (see [11, 22, 23]). Then we define a trilinear form:

$$J(\mathbf{u}, \mathbf{v}, \mathbf{w}) = \int_{\Xi} f\, \big|((\mathscr{R}\mathscr{P})(\mathbf{u}, \mathbf{v}))^+_N\big|\ |[\mathbf{w}_\tau]|\, d\xi, \quad \forall \mathbf{u}, \mathbf{v}, \mathbf{w} \in \mathscr{V}. \tag{13}$$

Moreover, we assume the following compatibility conditions:

$$\exists \mathbf{l} \in \mathscr{H} :\ (\mathbf{l}, \mathbf{w}) + a(\mathbf{v}_0, \mathbf{w}) + b(\mathbf{u}_0, \mathbf{w}) = \langle \mathbf{L}, \mathbf{w} \rangle, \quad \forall \mathbf{w} \in \mathscr{V}. \tag{14}$$

Let $\langle \cdot, \cdot \rangle_{-1/2,1/2}$ be the duality pairing between $H^{-1/2}(\Omega)$ and $H^{1/2}(\Omega)$. A variational formulation of problem P_0 is as follows (cf. [12, 16, 17]).

Problem P : Find $\mathbf{u} \in W^1_2(0, T; \mathscr{V}) \cap C^1([0, T]; H^{-1/2}(\Omega))$ such that $\mathbf{u}(t) \in \mathscr{K}$ for any $t \in (0, T)$, $\mathbf{v} \equiv \partial \mathbf{u}/\partial t$, and

$$
\begin{aligned}
&\langle \mathbf{v}(T), \mathbf{w}(T) - \mathbf{u}(T)\rangle_{-1/2,1/2} - (\mathbf{v}_0, \mathbf{w}(0) - \mathbf{u}_0)\\
&\quad + \int_0^T [-(\mathbf{v}, \partial_t \mathbf{w}) + a(\mathbf{v}, \mathbf{w}) + b(\mathbf{u}, \mathbf{w}) + J(\mathbf{u}, \mathbf{v}, \mathbf{w} + \mathbf{v} - \mathbf{u})]\, \mathrm{d}t\\
&\geq \int_0^T \langle \mathbf{L}, \mathbf{w} - \mathbf{u}\rangle\, \mathrm{d}t + \int_0^T \left[-|\mathbf{v}|^2 + a(\mathbf{v}, \mathbf{u}) + b(\mathbf{u}, \mathbf{u}) + J(\mathbf{u}, \mathbf{v}, \mathbf{v})\right] \mathrm{d}t, \quad (15)\\
&\quad \forall \mathbf{w} \in L_\infty(0, T : \mathscr{V}) \cap W_2^1(0, T; \mathscr{H}),\ \mathbf{w}(t) \in \mathscr{K}\ a.e.\ t \in (0, T).
\end{aligned}
$$

[N.B.: Problem P is formally equivalent to problem P_0.]

Our main theorem is as follows.

Theorem 1 *Problem P admits a solution.*

Let us rewrite problem P with the aid of the decomposition of Ω into Ω^+ and Ω^- stated above. To this end we introduce the following function spaces:

$$
\begin{aligned}
&\mathscr{H}^\pm = L_2(\Omega^\pm), \quad \mathscr{V}^\pm = \left\{\mathbf{v} \in H^1(\Omega^\pm) \mid \mathbf{v} = 0 \text{ a.e. on } \Gamma_U^\pm\right\},\\
&\hat{\mathscr{H}} = \mathscr{H}^+ \times \mathscr{H}^-, \quad \hat{\mathscr{V}} = \left\{\hat{\mathbf{v}} = (\mathbf{v}^+, \mathbf{v}^-) \in \mathscr{V}^+ \times \mathscr{V}^- \mid \mathbf{v}^+ = \mathbf{v}^- \text{ a.e. on } \Gamma_V\right\};
\end{aligned}
$$

the inner product $(\cdot, \cdot)_{\hat{\mathscr{H}}}$ and the associated norm $|\cdot|_{\hat{\mathscr{H}}}$ in $\hat{\mathscr{H}}$; the inner product $\langle\cdot, \cdot\rangle_{\hat{\mathscr{V}}}$ and the associated norm $|\cdot|_{\hat{\mathscr{V}}}$ in $\hat{\mathscr{V}}$. Note that $\mathbf{v} \in \mathscr{V}$ if and only if $\mathbf{v}^+ \in \mathscr{V}^+$ and $\mathbf{v}^- \in \mathscr{V}^-$ such that $\mathbf{v}^+ = \mathbf{v}^-$ a.e. on Γ_V. Further, define $\Psi : \mathscr{V} \to \hat{\mathscr{V}}$ by

$$
\mathbf{v} \mapsto \Psi \mathbf{v} =: \hat{\mathbf{v}} = (\mathbf{v}^+, \mathbf{v}^-). \tag{16}
$$

It is clear that Ψ is linear and continuous on $\mathscr{V}$ and $\|\Psi \mathbf{v}\|_{\hat{\mathscr{V}}} = \|\mathbf{v}\|$.

Now let $\hat{\mathscr{K}} = \Psi(\mathscr{K}) = \{\Psi \mathbf{v} \mid \mathbf{v} \in \mathscr{V},\ [v_N] \leq g \text{ a.e. on } \Xi\}$, and for any $\hat{\mathbf{v}} \in \hat{\mathscr{V}}$ we set $[\hat{v}_N] = [v_N]$ and $[\hat{\mathbf{v}}_\tau] = [\mathbf{v}_\tau]$, where $\mathbf{v} = \Psi^{-1}\hat{\mathbf{v}}$. Moreover, let

$$
\begin{aligned}
&H^s := W_2^s(\Omega^+) \times W_2^s(\Omega^-) \equiv H^s(\Omega^+) \times H^s(\Omega^-), \quad \forall s \in \mathbb{R},\\
&\langle \hat{\mathbf{u}}, \hat{\mathbf{v}}\rangle_{-s,s} = \langle \mathbf{u}^+, \mathbf{v}^+\rangle_{H^{-s}(\Omega^+), H^s(\Omega^+)} + \langle \mathbf{u}^-, \mathbf{v}^-\rangle_{H^{-s}(\Omega^-), H^s(\Omega^-)}, \quad \forall \hat{\mathbf{u}} \in H^{-s}, \hat{\mathbf{v}} \in H^s,
\end{aligned}
$$

and

$$
\begin{cases}
\hat{a}(\hat{\mathbf{v}}, \hat{\mathbf{w}}) = a(\mathbf{v}, \mathbf{w}) = a^+(\mathbf{v}^+, \mathbf{w}^+) + a^-(\mathbf{v}^-, \mathbf{w}^-),\\
\hat{b}(\hat{\mathbf{u}}, \hat{\mathbf{w}}) = b(\mathbf{u}, \mathbf{w}) = b^+(\mathbf{u}^+, \mathbf{w}^+) + b^-(\mathbf{u}^-, \mathbf{w}^-),\\
\hat{J}(\hat{\mathbf{u}}, \hat{\mathbf{v}}, \hat{\mathbf{w}}) = \displaystyle\int_\Xi f \left|((\mathscr{R}\mathscr{P})(\mathbf{u}, \mathbf{v}))_N^+\right| |[\mathbf{w}_\tau]|\, \mathrm{d}\xi,\\
\langle \hat{\mathbf{L}}, \hat{\mathbf{w}}\rangle_{\hat{\mathscr{V}}} = \langle \mathbf{L}, \mathbf{w}\rangle, \quad (\hat{\mathbf{l}}, \hat{\mathbf{w}})_{\hat{\mathscr{H}}} = (\mathbf{l}, \mathbf{w}),\\
\quad \forall \hat{\mathbf{u}} = (\mathbf{u}^+, \mathbf{u}^-),\ \hat{\mathbf{v}} = (\mathbf{v}^+, \mathbf{v}^-),\ \hat{\mathbf{w}} = (\mathbf{w}^+, \mathbf{w}^-) \in \hat{\mathscr{V}}\\
\qquad \text{satisfying } \Psi \mathbf{u} = \hat{\mathbf{u}},\ \Psi \mathbf{v} = \hat{\mathbf{v}},\ \Psi \mathbf{w} = \hat{\mathbf{w}}.
\end{cases} \tag{17}
$$

Here

$$a^{\pm}(\mathbf{v}^{\pm}, \mathbf{w}^{\pm}) = \int_{\Omega^{\pm}} \big(2\mu \mathscr{D}(\mathbf{v}^{\pm}) : \mathscr{D}(\mathbf{w}^{\pm}) + \mu' \operatorname{tr}(\mathscr{D}(\mathbf{v}^{\pm})) \operatorname{tr}(\mathscr{D}(\mathbf{w}^{\pm}))\big)\, \mathrm{d}x,$$

$$b^{\pm}(\mathbf{u}^{\pm}, \mathbf{w}^{\pm}) = \int_{\Omega^{\pm}} \big(2\lambda \mathscr{D}(\mathbf{u}^{\pm}) : \mathscr{D}(\mathbf{w}^{\pm}) + \lambda' \operatorname{tr}(\mathscr{D}(\mathbf{u}^{\pm})) \operatorname{tr}(\mathscr{D}(\mathbf{w}^{\pm}))\big)\, \mathrm{d}x.$$

The compatibility condition (14) can be written as

$$\exists \hat{\mathbf{l}} \in \hat{\mathscr{H}} : \; (\hat{\mathbf{l}}, \hat{\mathbf{w}})_{\hat{\mathscr{H}}} + \hat{a}(\hat{\mathbf{v}}_0, \hat{\mathbf{w}}) + \hat{b}(\hat{\mathbf{u}}_0, \hat{\mathbf{w}}) = \langle \hat{\mathbf{L}}, \hat{\mathbf{w}} \rangle_{\hat{\mathscr{V}}}, \quad \forall \hat{\mathbf{w}} \in \hat{\mathscr{V}}. \tag{18}$$

Now, we consider the following auxiliary problem.
Problem $\hat{P}$: Find $\hat{\mathbf{u}} = (\mathbf{u}^+, \mathbf{u}^-) \in W_2^1(0, T; \hat{\mathscr{V}}) \cap C^1([0, T]; H^{-1/2})$ such that $\hat{\mathbf{u}}(t) \in \hat{\mathscr{K}}$ for any $t \in (0, T)$, $\hat{\mathbf{v}} \equiv \partial \hat{\mathbf{u}} / \partial t$, and

$$\begin{aligned}
&\langle \hat{\mathbf{v}}(T), \hat{\mathbf{w}}(T) - \hat{\mathbf{u}}(T) \rangle_{-1/2,1/2} - (\hat{\mathbf{v}}_0, \hat{\mathbf{w}}(0) - \hat{\mathbf{u}}_0)_{\hat{\mathscr{H}}} \\
&+ \int_0^T \Big[-(\hat{\mathbf{v}}, \partial_t \hat{\mathbf{w}})_{\hat{\mathscr{H}}} + \hat{a}(\hat{\mathbf{v}}, \hat{\mathbf{w}}) + \hat{b}(\hat{\mathbf{u}}, \hat{\mathbf{w}}) + \hat{J}(\hat{\mathbf{u}}, \hat{\mathbf{v}}, \hat{\mathbf{w}} + \hat{\mathbf{v}} - \hat{\mathbf{u}}) \Big]\, \mathrm{d}t \\
&\geq \int_0^T \langle \hat{\mathbf{L}}, \hat{\mathbf{w}} - \hat{\mathbf{u}} \rangle_{\hat{\mathscr{V}}}\, \mathrm{d}t + \int_0^T \Big[-|\hat{\mathbf{v}}|^2_{\hat{\mathscr{H}}} + \hat{a}(\hat{\mathbf{v}}, \hat{\mathbf{u}}) + \hat{b}(\hat{\mathbf{u}}, \hat{\mathbf{u}}) + \hat{J}(\hat{\mathbf{u}}, \hat{\mathbf{v}}, \hat{\mathbf{v}}) \Big]\, \mathrm{d}t, \\
&\qquad \forall \hat{\mathbf{w}} \in L_\infty(0, T; \hat{\mathscr{V}}) \cap W_2^1(0, T; \hat{\mathscr{H}}), \; \hat{\mathbf{w}}(t) \in \hat{\mathscr{K}} \; a.e.\, t \in (0, T).
\end{aligned} \tag{19}$$

It is easy to check the following equivalence result.

Proposition 1 *Under the above assumptions, problems P and $\hat{P}$ are equivalent in the following sense:*

(i) If $\mathbf{u}$ is a solution of problem P, then $\Psi \mathbf{u} = (\mathbf{u}^+, \mathbf{u}^-)$ is a solution of problem $\hat{P}$.
(ii) Conversely, if $\hat{\mathbf{u}} = (\mathbf{u}^+, \mathbf{u}^-)$ is a solution of problem $\hat{P}$, then $\mathbf{u} \in \mathscr{V}$, with $\mathbf{u} = \mathbf{u}^+$ a.e. in Ω^+ and $\mathbf{u} = \mathbf{u}^-$ a.e. in Ω^-, is a solution of problem P.

Note that a solution to problem P does not depend on the choice of Γ_V, and (19) implies $\mathscr{P}^+\mathbf{n}^+ = -\mathscr{P}^-\mathbf{n}^-$ on Γ_V.
We shall solve problem P by virtue of a penalty method (see [12, 16]).

3 Penalized Problem

In this section, we consider a penalized problem consisting of the same equations in Ω and the same boundary conditions on Γ_U, Γ_F, Γ mentioned above except for the unilateral contact conditions, whose solution is denoted by $\mathbf{u}_\varepsilon$, $\mathbf{v}_\varepsilon := \partial \mathbf{u}_\varepsilon / \partial t$. The penetration of the crack faces is penalized.

3.1 Formulation of Penalized Problem

For almost every $t \in (0, T)$, the contact conditions on Ξ are penalized as follows:

$$\begin{cases} (\mathscr{P})_N^+(\mathbf{u}_\varepsilon, \mathbf{v}_\varepsilon) = (\mathscr{P})_N^-(\mathbf{u}_\varepsilon, \mathbf{v}_\varepsilon) = -\dfrac{1}{\varepsilon}\left([u_{\varepsilon N}] - g\right)_+ \quad \text{on } \Xi \\ \qquad [(\cdot)_+ := \max(0, \cdot)], \\ \mathscr{P}_\tau^+(\mathbf{u}_\varepsilon, \mathbf{v}_\varepsilon) = -\mathscr{P}_\tau^-(\mathbf{u}_\varepsilon, \mathbf{v}_\varepsilon), \quad \left|\mathscr{P}_\tau^+(\mathbf{u}_\varepsilon, \mathbf{v}_\varepsilon)\right| \le f\left|((\mathscr{R}\mathscr{P})(\mathbf{u}_\varepsilon, \mathbf{v}_\varepsilon))_N^+\right|, \\ [\mathbf{v}_{\varepsilon\tau}] = 0 \quad \text{when } \left|\mathscr{P}_\tau^+(\mathbf{u}_\varepsilon, \mathbf{v}_\varepsilon)\right| < f\left|((\mathscr{R}\mathscr{P})(\mathbf{u}_\varepsilon, \mathbf{v}_\varepsilon))_N^+\right|, \\ \exists\, \lambda \ge 0 \text{ such that } [\mathbf{v}_{\varepsilon\tau}] = -\lambda \mathscr{P}_\tau^+(\mathbf{u}_\varepsilon, \mathbf{v}_\varepsilon) \\ \qquad \text{when } \left|\mathscr{P}_\tau^+(\mathbf{u}_\varepsilon, \mathbf{v}_\varepsilon)\right| = f\left|((\mathscr{R}\mathscr{P})(\mathbf{u}_\varepsilon, \mathbf{v}_\varepsilon))_N^+\right|. \end{cases} \tag{20}$$

Let the mapping $\Phi_\varepsilon : \mathscr{V} \times \mathscr{V} \to \mathbb{R}$ be defined by

$$\Phi_\varepsilon(\mathbf{u}, \mathbf{w}) = \frac{1}{\varepsilon}\int_\Xi ([u_N] - g)_+ [w_N]\, d\xi, \quad \forall \mathbf{u},\ \mathbf{w} \in \mathscr{V}.$$

A variational formulation of the problem corresponding to (20) is as follows:

Problem P_ε: Find $\mathbf{u}_\varepsilon \in W_2^1(0, T; \mathscr{V}) \cap W_2^2(0, T; \mathscr{H})$, $\mathbf{v}_\varepsilon \equiv \partial \mathbf{u}_\varepsilon/\partial t$, such that $\mathbf{u}_\varepsilon(0) = \mathbf{u}_0$, $\mathbf{v}_\varepsilon(0) = \mathbf{v}_0$ and

$$\begin{aligned} &\left(\frac{\partial \mathbf{v}_\varepsilon}{\partial t}, \mathbf{w} - \mathbf{v}_\varepsilon\right) + a(\mathbf{v}_\varepsilon, \mathbf{w} - \mathbf{v}_\varepsilon) + b(\mathbf{u}_\varepsilon, \mathbf{w} - \mathbf{v}_\varepsilon) + \Phi_\varepsilon(\mathbf{u}_\varepsilon, \mathbf{w} - \mathbf{v}_\varepsilon) \\ &+ J(\mathbf{u}_\varepsilon, \mathbf{v}_\varepsilon, \mathbf{w}) - J(\mathbf{u}_\varepsilon, \mathbf{v}_\varepsilon, \mathbf{v}_\varepsilon) \ge \langle \mathbf{L}, \mathbf{w} - \mathbf{v}_\varepsilon \rangle, \quad \forall \mathbf{w} \in \mathscr{V}, \text{ a.e. } t \in (0, T). \end{aligned} \tag{21}$$

To solve problem P_ε, let us consider an abstract problem P^* described in Sect. 3.2.

3.2 Abstract Evolution Problem

Let $(\mathscr{H}, |\cdot|)$ and $(\mathscr{V}, \|\cdot\|)$ be two Hilbert spaces with respective inner products denoted by $(\cdot, \cdot)$ and $\langle \cdot, \cdot \rangle$ such that $\mathscr{V}$ is dense in $\mathscr{H}$ and the embedding from $\mathscr{V}$ into $\mathscr{H}$ is compact. We also introduce the space $\mathscr{W}$ and the set $\mathscr{X}$ as

$$\begin{aligned} \mathscr{W} &= W_2^2(0, T; \mathscr{H}) \cap W_2^1(0, T; \mathscr{V}), \\ \mathscr{X} &= \left\{ w \in \mathscr{W} \;\middle|\; \left(w, \frac{\partial w}{\partial t}\right)(0) = (w_0, w_1) \right\}, \end{aligned}$$

where $w_0,\ w_1 \in \mathscr{V}$. Let a and b be two bilinear, symmetric, continuous, and $\mathscr{V}$-elliptic forms defined on $\mathscr{V} \times \mathscr{V}$, in the following sense:

$$\begin{cases} \exists M_a,\ M_b > 0 : a(u, w) \leq M_a\|u\|\|w\|,\ b(u, w) \leq M_b\|u\|\|w\|,\ \ \forall u,\ w \in \mathscr{V}, \\ \exists m_a,\ m_b > 0 : a(u, u) \geq m_a\|u\|^2,\ b(u, u) \geq m_b\|u\|^2,\ \ \forall u \in \mathscr{V}. \end{cases} \tag{22}$$

Let $\beta : \mathscr{V} \to \mathbb{R}$ and $\phi : [0, T] \times \mathscr{V} \times \mathscr{V} \times \mathscr{V} \to \mathbb{R}$ be two weakly continuous mappings such that

$$\begin{cases} \phi(t, u, v, w_1 + w_2) \leq \phi(t, u, v, w_1) + \phi(t, u, v, w_2), \\ \phi(t, u, v, \theta w) = \theta\phi(t, u, v, w),\ \ \forall t \in [0, T],\ \forall u,\ v,\ w_1,\ w_2 \in \mathscr{V},\ \theta \geq 0, \\ \phi(\cdot, 0, 0, \cdot) = 0,\quad \beta(0) = 0, \end{cases} \tag{23}$$

$$\begin{cases} \exists \eta_0 > 0 : \forall t_1,\ t_2 \in [0, T],\ \forall u_1,\ u_2,\ v_1,\ v_2,\ w \in \mathscr{V}, \\ \quad |\phi(t_1, u_1, v_1, w) - \phi(t_2, u_2, v_2, w)| \\ \qquad \leq \eta_0\,(|t_1 - t_2| + |\beta(u_1 - u_2)| + |v_1 - v_2|)\,\|w\|, \end{cases} \tag{24}$$

$$\begin{cases} \exists \eta > 0 : \forall t_1,\ t_2 \in [0, T],\ \forall u_1,\ u_2,\ v_1,\ v_2,\ w_1,\ w_2 \in \mathscr{V}, \\ \quad |\phi(t_1, u_1, v_1, w_1) - \phi(t_1, u_1, v_1, w_2) + \phi(t_2, u_2, v_2, w_2) - \phi(t_2, u_2, v_2, w_1)| \\ \qquad \leq \eta\,(|t_1 - t_2| + \|u_1 - u_2\| + |v_1 - v_2|)\,\|w_1 - w_2\|. \end{cases} \tag{25}$$

Let $L \in W^1_\infty(0, T; \mathscr{V})$, and let C_L be the Lipschitz constant of L. Assume the compatibility condition on the initial data,

$$\exists l \in \mathscr{H} : (l, w) + a(u_1, w) + b(u_0, w) + \phi(0, u_0, u_1, w) = \langle L(0), w\rangle,\ \ \forall w \in \mathscr{V}, \tag{26}$$

and consider the problem,

Problem P^*: Find $u \in \mathscr{X}$ such that for almost all $t \in (0, T)$

$$\begin{aligned} &(\partial_t^2 u, w - \partial_t u) + a(\partial_t u, w - \partial_t u) + b(u, w - \partial_t u) + \phi(t, u, \partial_t u, w) \\ &\quad -\phi(t, u, \partial_t u, \partial_t u) \geq \langle L, w - \partial_t u\rangle,\ \ \forall w \in \mathscr{V}\ \ (\partial_t := \partial/\partial t). \end{aligned} \tag{27}$$

We shall prove the existence of a solution to problem P^* by an incremental technique due to Ricaud and Pratt [28]. However, its proof is much the same as that in [9], so that here we describe in outline only.

For any $n = 2, 3, 4, \ldots,\ j = 1, 2, \ldots, n - 1$ set

$$\begin{gathered} \Delta t = T/n,\quad t^j = j\Delta t,\quad L^j = L(t^j),\quad d^j = \frac{u^{j+1} - u^{j-1}}{2\Delta t}, \\ \delta^j = \frac{u^{j+1} - u^j}{\Delta t},\quad \gamma^j = \frac{\delta^j - \delta^{j-1}}{\Delta t} = \frac{u^{j+1} - 2u^j + u^{j-1}}{(\Delta t)^2}, \\ u^0 = u_0,\quad \delta^0 = u_1,\quad u^1 = u_0 + \Delta t u_1. \end{gathered}$$

We consider the following incremental problems for $j = 1, 2, \ldots, n - 1$.

Problem P_n^{j+1}: Find $u^{j+1} \in \mathscr{V}$ such that

$$(\gamma^j, w - d^j) + a(d^j, w - d^j) + b(u^{j+1}, w - d^j) + \phi(t^j, u^{j+1}, d^j, w) \\ -\phi(t^j, u^{j+1}, d^j, d^j) \geq \langle L^j, w - d^j \rangle, \quad \forall w \in \mathscr{V}. \tag{28}$$

Then, we prove the following assertions in order (cf. [2, 9, 14, 28]).

1. Each problem P_n^{j+1} has a unique solution u^{j+1} if Δt is sufficiently small.
2. If the compatibility condition (26) holds and if Δt is sufficiently small, then the term γ^1 is bounded in $\mathscr{H}$ and the term $\Delta t \gamma^1$ is bounded in $\mathscr{V}$ independently of Δt.

Assertion 2 yields that d^1 and δ^1 are bounded in $\mathscr{V}$ independently of Δt if Δt is sufficiently small.

3. If Δt is sufficiently small, then γ^j are bounded in $\mathscr{H}$ and δ^j are bounded in $\mathscr{V}$ independent of Δt for all $j = 2, 3, \ldots, n-1$.

Now we define the sequence $\{u_n\}_{n=2,3,4,\ldots}$:

$$u_n(t) = u^1, \quad \bar{u}_n(t) = \tilde{u}_n(t) = \frac{3u^0 - u^1}{2} + t\delta^0, \quad t \in [0, t^1],$$

$$\begin{cases} u_n(t) = u^{j+1}, \quad \bar{u}_n(t) = \dfrac{u^{j-1} + u^j}{2} + (t - t^j)\, d^j, \\ \tilde{u}_n(t) = \dfrac{u^{j-1} + u^j}{2} + (t - t^j)\, \delta^{j-1} + \dfrac{(t - t^j)^2}{2} \gamma^j, \quad t \in (t^j, t^{j+1}] \end{cases}$$

$(j = 1, 2, \ldots, n-1)$.
It is clear that $u_n \in L_2(0, T; \mathscr{V})$, $\bar{u}_n \in W_2^1(0, T; \mathscr{V})$, $\tilde{u}_n \in W_2^2(0, T; \mathscr{H})$.

The sequence of problems $\{P_n^{j+1}\}_{j=1,2,\ldots,n-1}$ is equivalent to

$$\int_0^T \Big(\big(\partial_t^2 \tilde{u}_n, w - \partial_t \bar{u}_n \big) + a(\partial_t \bar{u}_n, w - \partial_t \bar{u}_n) + b(u_n, w - \partial_t \bar{u}_n) \\ + \phi_n(t, u_n, \partial_t \bar{u}_n, w) - \phi_n(t, u_n, \partial_t \bar{u}_n, \partial_t \bar{u}_n) \Big) \, \mathrm{d}t \geq \int_0^T \langle L_n, w - \partial_t \bar{u}_n \rangle \tag{29}$$

for $\forall w \in L_2(0, T; \mathscr{V})$, where $\phi_n(t, u, v, w) = \phi(t^j, u, v, w)$, $L_n(t) = L^j$ for any $t \in (t^j, t^{j+1}]$ $(j = 1, 2, \ldots, n-1)$, $\phi_n(t, u, v, w) = \phi(0, u, v, w)$, $L_n(t) = L^0$ for any $t \in [0, t^1]$.

The estimates in assertions 2 and 3 imply that there exist u, $\bar{u}$, $\tilde{u}$ such that, up to a subsequence, $u_n \overset{w}{\rightharpoonup} u$ in $L_2(0, T; \mathscr{V})$, $\bar{u}_n \overset{w}{\rightharpoonup} \bar{u}$ in $W_2^1(0, T; \mathscr{V})$ and $\tilde{u}_n \overset{w}{\rightharpoonup} \tilde{u}$ in $W_2^2(0, T; \mathscr{H})$ as $n \to 0$.

4. The limits u, $\bar{u}$, and $\tilde{u}$ are equal, i.e.,

$$\begin{cases} \lim\limits_{n\to\infty} \|\bar{u}_n - u_n\|_{L_2(0,T;\mathscr{V})} = 0, \quad \lim\limits_{n\to\infty} \|\bar{u}_n - \tilde{u}_n\|_{L_2(0,T;\mathscr{V})} = 0, \\ \lim\limits_{n\to\infty} \|\partial_t\bar{u}_n - \partial_t\tilde{u}_n\|_{L_2(0,T;\mathscr{H})} = 0. \end{cases} \tag{30}$$

Since $u_n(0) = u_0 + \Delta t u_1$, $\bar{u}_n(0) = \tilde{u}_n(0) = u_0 - \Delta t u_1/2$ and $\partial_t\bar{u}_n(0) = \partial_t\tilde{u}_n(0) = u_1$, u satisfies the initial conditions. The facts $\{\bar{u}_n(t)\}_n \subset W_2^1(0, T; \mathscr{V})$ and $\{\tilde{u}_n(t)\}_n \subset \mathscr{W}$ imply that, by a diagonal process, up to a subsequence, for any $t \in [0, T]$

$$u_n(t) \overset{w}{\rightharpoonup} u(t)\, in\, \mathscr{V},\ \ \tilde{u}_n(t) \overset{w}{\rightharpoonup} u(t)\, in\, \mathscr{V},\ \ \partial_t\tilde{u}_n(t) \overset{w}{\rightharpoonup} \partial_t u(t)\, in\, \mathscr{H}\ \text{as}\, n \to \infty. \tag{31}$$

Now we discuss the passing to the limit of the nonlinear terms. Since

$$\int_0^T (\partial_t^2\tilde{u}_n(t), \partial_t\bar{u}_n(t))\,\mathrm{d}t = \sum_{i=1}^{n-1} \Delta t(\gamma^i, d^i) = \frac{1}{2}\left(|\partial_t\tilde{u}_n(T)|^2 - |\partial_t\tilde{u}_n(0)|^2\right)$$

by (31), we arrive at

$$\begin{aligned} \liminf_{n\to\infty} \int_0^T (\partial_t^2\tilde{u}_n(t), \partial_t\bar{u}_n(t))\,\mathrm{d}t &\geq \frac{1}{2}\left(|\partial_t u(T)|^2 - |u_1(0)|^2\right) \\ &= \int_0^T (\partial_t^2 u(t), \partial_t u(t))\,\mathrm{d}t. \end{aligned} \tag{32}$$

It is easily seen that the following relation holds:

$$\begin{aligned} \int_0^T b(u_n, \partial_t\bar{u}_n)\,\mathrm{d}t &= \int_0^{t^1} b(u_n, \partial_t\bar{u}_n)\,\mathrm{d}t + \sum_{i=1}^{n-1} \int_{t^i}^{t^{i+1}} b(u_n, \partial_t\bar{u}_n)\,\mathrm{d}t \\ &= \Delta t\, b(u^1, \delta^0) + \Delta t \sum_{i=1}^{n-1} b\left(u^{i+1}, \frac{u^{i+1} - u^{i-1}}{2\Delta t}\right) \\ &\geq \frac{1}{2}\, b(u^n, u^n) - \frac{1}{2}\, b(u^1, u^1) - \frac{\Delta t}{2}\left(b(u^n, \delta^{n-1}) - 3b(u^1, \delta^0)\right) \\ &\quad + \frac{(\Delta t)^2}{4}\left(b(\delta^{n-1}, \delta^{n-1}) - b(\delta^0, \delta^0)\right). \end{aligned}$$

This, together with (31) and $u_n(T) = u^n$, leads to

$$\liminf_{n\to\infty}\int_0^T b(u_n,\partial_t\bar{u}_n)\,\mathrm{d}t \geq \liminf_{n\to\infty}\frac{1}{2}\left(b(u_n(T),u_n(T)) - b(u_n(0),u_n(0))\right)$$

$$\geq \frac{1}{2}\left(b(u(T),u(T)) - b(u_0,u_0)\right) = \int_0^T b(u,\partial_t u)\,\mathrm{d}t. \tag{33}$$

By virtue of the lower semi-continuity of a we also have

$$\liminf_{n\to\infty}\int_0^T a(\partial_t\bar{u}_n,\partial_t\bar{u}_n)\,\mathrm{d}t \geq \int_0^T a(\partial_t u,\partial_t u)\,\mathrm{d}t. \tag{34}$$

5.

$$\liminf_{n\to\infty}\int_0^T \phi_n(t,u_n,\partial_t\bar{u}_n,\partial_t\bar{u}_n)\,\mathrm{d}t \geq \int_0^T \phi(t,u,\partial_t u,\partial_t u)\,\mathrm{d}t. \tag{35}$$

6.

$$\lim_{n\to\infty}\int_0^T \phi_n(t,u_n,\partial_t\bar{u}_n,w)\,\mathrm{d}t \geq \int_0^T \phi(t,u,\partial_t u,w)\,\mathrm{d}t, \quad \forall w\in L_2(0,T;\mathscr{V}). \tag{36}$$

In proving (36), we make use of Theorem 2 due to Simon with $\mathscr{F}=\{\partial_t\tilde{u}_n\}_n$, $X=\mathscr{V}$, $U=\mathscr{H}$, $Y=\mathscr{H}$, $p=2$, which guarantees that, up to a subsequence, $\{\partial_t\tilde{u}_n\}_n$ strongly converges to $\partial_t u$ in $L_2(0,T;\mathscr{H})$.

Theorem 2 (Simon [30]) *Suppose that X, U, and Y be three Banach spaces such that $X\subset U\subset Y$ with compact embedding $X\to U$. Let $\mathscr{F}$ be bounded in $L_p(0,T;X)$, $1\leq p<\infty$, and $\partial\mathscr{F}/\partial t=\{\partial_t f \mid f\in\mathscr{F}\}$ be bounded in $L_1(0,T;Y)$. Then $\mathscr{F}$ is relatively compact in $L_p(0,T;U)$.*

Let $\mathscr{F}$ be bounded in $L_1(0,T;X)$ and $\partial\mathscr{F}/\partial t$ be bounded in $L_r(0,T;Y)$, $r>1$. Then $\mathscr{F}$ is relatively compact in $C([0,T];U)$.

7. Problem P^* admits at most one solution.

The above seven assertions imply the following.

Theorem 3 *Under the assumptions on u_0, u_1, L, (22)–(25) and the compatibility condition (26), there exists a unique solution to problem P^*.*

Indeed, by using (32)–(35) and (36), we can pass to the limit in (29). Thus, $u\in\mathscr{X}$ is the solution to the following problem:

$$\int_0^T \left((\partial_t^2 u, w - \partial_t u) + a(\partial_t u, w - \partial_t u) + b(u, w - \partial_t u) + \phi(t, u, \partial_t u, w)\right.$$

$$\left. -\phi(t, u, \partial_t u, \partial_t u)\right) \, \mathrm{d}t \geq \int_0^T \langle L, w - \partial_t u\rangle \, \mathrm{d}t, \quad \forall w \in L_2(0, T; \mathscr{V}). \quad (37)$$

Lebesgue's theorem implies that u is a solution to problem P^*. Uniqueness result comes from the fact in assertion 7.

3.3 Unique Solvability of Problem P_ε

Let $\hat{\Phi}_\varepsilon$ be a form defined by

$$\hat{\Phi}_\varepsilon(\hat{\mathbf{u}}, \hat{\mathbf{v}}) = \Phi_\varepsilon(\Psi^{-1}\hat{\mathbf{u}}, \Psi^{-1}\hat{\mathbf{v}}), \quad \forall \hat{\mathbf{u}}, \ \hat{\mathbf{v}} \in \hat{\mathscr{V}},$$

and consider an auxiliary penalized problem equivalent to problem P_ε, corresponding to the decomposition of Ω used in Sect. 2.

Problem $\hat{P}_\varepsilon$: Find $\hat{\mathbf{u}}_\varepsilon = (\mathbf{u}_\varepsilon^+, \mathbf{u}_\varepsilon^-) \in W_2^1(0, T; \hat{\mathscr{V}}) \cap W_2^2(0, T; \hat{\mathscr{H}})$, $\hat{\mathbf{v}}_\varepsilon \equiv \partial\hat{\mathbf{u}}_\varepsilon/\partial t$, such that $\hat{\mathbf{u}}_\varepsilon(0) = \hat{\mathbf{u}}_0$, $\hat{\mathbf{v}}_\varepsilon(0) = \hat{\mathbf{v}}_0$ and

$$(\partial_t \hat{\mathbf{v}}_\varepsilon, \hat{\mathbf{w}} - \hat{\mathbf{v}}_\varepsilon)_{\hat{\mathscr{H}}} + \hat{a}(\hat{\mathbf{v}}_\varepsilon, \hat{\mathbf{w}} - \hat{\mathbf{v}}_\varepsilon) + \hat{b}(\hat{\mathbf{u}}_\varepsilon, \hat{\mathbf{w}} - \hat{\mathbf{v}}_\varepsilon) + \hat{\Phi}_\varepsilon(\hat{\mathbf{u}}_\varepsilon, \hat{\mathbf{w}} - \hat{\mathbf{v}}_\varepsilon)$$
$$+\hat{J}(\hat{\mathbf{u}}_\varepsilon, \hat{\mathbf{v}}_\varepsilon, \hat{\mathbf{w}}) - \hat{J}(\hat{\mathbf{u}}_\varepsilon, \hat{\mathbf{v}}_\varepsilon, \hat{\mathbf{v}}_\varepsilon) \geq \langle \hat{\mathbf{L}}, \hat{\mathbf{w}} - \hat{\mathbf{v}}_\varepsilon\rangle_{\hat{\mathscr{V}}}, \quad \forall \hat{\mathbf{w}} \in \hat{\mathscr{V}}, \text{ a.e. } t \in (0, T). \quad (38)$$

It is easily seen that the following equivalence result holds.

Proposition 2 *Problems P_ε and $\hat{P}_\varepsilon$ are equivalent in the following sense:*

(i) *If $\mathbf{u}_\varepsilon$ is a solution to problem P_ε, then $\hat{\mathbf{u}}_\varepsilon = (\mathbf{u}_\varepsilon^+, \mathbf{u}_\varepsilon^-)$ is a solution to problem $\hat{P}_\varepsilon$, where $\mathbf{u}_\varepsilon^+$ and $\mathbf{u}_\varepsilon^-$ are the restrictions of $\mathbf{u}_\varepsilon$ on Ω^+ and Ω^-, respectively.*
(ii) *Conversely, if $\hat{\mathbf{u}}_\varepsilon = (\mathbf{u}_\varepsilon^+, \mathbf{u}_\varepsilon^-)$ is a solution to problem $\hat{P}_\varepsilon$, then $\mathbf{u}_\varepsilon \in \mathscr{V}$, with $\mathbf{u}_\varepsilon = \mathbf{u}_\varepsilon^+$ a.e. in Ω^+ and $\mathbf{u}_\varepsilon = \mathbf{u}_\varepsilon^-$ a.e. in Ω^-, is a solution to problem P_ε.*

Theorem 4 *Under the assumptions of Sect. 2, problems P_ε and $\hat{P}_\varepsilon$ admit unique solutions.*

Proof Apply Theorem 3 to $H = \hat{\mathscr{H}}$, $V = \hat{\mathscr{V}}$, $u_0 = \hat{\mathbf{u}}_0$, $u_1 = \hat{\mathbf{v}}_0$, $a = \hat{a}$, $b = \hat{b}$, $L = \hat{\mathbf{L}}$, $\phi(\hat{\mathbf{u}}, \hat{\mathbf{v}}, \hat{\mathbf{w}}) = \hat{\Phi}_\varepsilon(\hat{\mathbf{u}}, \hat{\mathbf{w}}) + \hat{J}(\hat{\mathbf{u}}, \hat{\mathbf{v}}, \hat{\mathbf{w}})$, $\forall \hat{\mathbf{u}}, \ \hat{\mathbf{v}}, \ \hat{\mathbf{w}} \in \hat{\mathscr{V}}$. By using meas$(\Gamma_U^\pm) > 0$ and the Lipschitz continuity of the boundaries $\partial\Omega^\pm$, Korn's inequality implies that there exist $m_a^\pm > 0$ and $m_b^\pm > 0$ such that $a^\pm(\mathbf{v}, \mathbf{v}) \geq m_a^\pm \|\mathbf{v}\|_{\mathscr{V}^\pm}^2$, $b^\pm(\mathbf{v}, \mathbf{v}) \geq m_b^\pm \|\mathbf{v}\|_{\mathscr{V}^\pm}^2$ for any $\mathbf{v} \in \mathscr{V}^\pm$, and we get

$$\begin{cases} \hat{a}(\hat{\mathbf{v}}, \hat{\mathbf{v}}) \geq \hat{m}_a \, \|\hat{\mathbf{v}}\|^2_{\hat{\mathscr{V}}}, & \forall \hat{\mathbf{v}} \in \hat{\mathscr{V}}, \quad \text{where } \hat{m}_a = \min(m_a^+, m_a^-), \\ \hat{b}(\hat{\mathbf{v}}, \hat{\mathbf{v}}) \geq \hat{m}_b \, \|\hat{\mathbf{v}}\|^2_{\hat{\mathscr{V}}}, & \forall \hat{\mathbf{v}} \in \hat{\mathscr{V}}, \quad \text{where } \hat{m}_b = \min(m_b^+, m_b^-). \end{cases} \tag{39}$$

For any $\hat{\mathbf{u}}, \hat{\mathbf{v}} \in \hat{\mathscr{V}}$, $\hat{\Phi}_\varepsilon(\hat{\mathbf{u}}, \cdot)$ is linear on $\hat{\mathscr{V}}$ and $\hat{J}(\hat{\mathbf{u}}, \hat{\mathbf{v}}, \cdot)$ is a semi-norm on $\hat{\mathscr{V}}$, so that ϕ satisfies (23). Let us define β by $\beta(\hat{\mathbf{u}}) := \|[\hat{u}_N]\|_{L_2(\Xi)} + |\hat{\mathbf{u}}|_{\hat{\mathscr{H}}}$, $\forall \hat{\mathbf{u}} \in \hat{\mathscr{V}}$, which is clearly weakly continuous on $\hat{\mathscr{V}}$. The form ϕ satisfies (23) and (24) due to the averaging operator $\mathscr{R}$ (see (12)). Since the function $s \mapsto (s-g)_+$ is Lipschitz continuous on $\mathbb{R}$, we have the following inequality for $\hat{\Phi}_\varepsilon$:

$$\begin{aligned} \exists\, \eta_\varepsilon > 0 : \; & \left| \hat{\Phi}_\varepsilon(\hat{\mathbf{u}}_1, \hat{\mathbf{w}}_1) - \hat{\Phi}_\varepsilon(\hat{\mathbf{u}}_1, \hat{\mathbf{w}}_2) + \hat{\Phi}_\varepsilon(\hat{\mathbf{u}}_2, \hat{\mathbf{w}}_2) - \hat{\Phi}_\varepsilon(\hat{\mathbf{u}}_2, \hat{\mathbf{w}}_1) \right| \\ & \leq \eta_\varepsilon \|\hat{\mathbf{u}}_1 - \hat{\mathbf{u}}_2\|_{\hat{\mathscr{V}}} \, \|\hat{\mathbf{w}}_1 - \hat{\mathbf{w}}_2\|_{\hat{\mathscr{V}}}, \quad \forall \hat{\mathbf{u}}_1, \hat{\mathbf{u}}_2, \, \hat{\mathbf{w}}_1, \, \hat{\mathbf{w}}_2 \in \hat{\mathscr{V}}. \end{aligned} \tag{40}$$

Inequality (12) yields the estimate:

$$\begin{aligned} \exists\, \eta' > 0 : \; & \left| \hat{J}(\hat{\mathbf{u}}_1, \hat{\mathbf{v}}_1, \hat{\mathbf{w}}_1) - \hat{J}(\hat{\mathbf{u}}_1, \hat{\mathbf{v}}_1, \hat{\mathbf{w}}_2) + \hat{J}(\hat{\mathbf{u}}_2, \hat{\mathbf{v}}_2, \hat{\mathbf{w}}_2) - \hat{J}(\hat{\mathbf{u}}_2, \hat{\mathbf{v}}_2, \hat{\mathbf{w}}_1) \right| \\ & \leq \eta' \left(\left|\hat{\mathbf{u}}_2 - \hat{\mathbf{u}}_1\right|_{\hat{\mathscr{H}}} + \left|\hat{\mathbf{v}}_2 - \hat{\mathbf{v}}_1\right|_{\hat{\mathscr{H}}} \right) \left\|\hat{\mathbf{w}}_2 - \hat{\mathbf{w}}_1\right\|_{\hat{\mathscr{V}}}, \\ & \forall \hat{\mathbf{u}}_1, \hat{\mathbf{u}}_2, \, \hat{\mathbf{v}}_1 \, \hat{\mathbf{v}}_2, \, \hat{\mathbf{w}}_1, \, \hat{\mathbf{w}}_2 \in \hat{\mathscr{V}}. \end{aligned} \tag{41}$$

Relations (40)–(41) and (18) lead to (25) and (26), respectively. Hence, problem $\hat{P}_\varepsilon$ has a unique solution $\hat{\mathbf{u}}_\varepsilon \in W_2^1(0, T; \hat{\mathscr{V}}) \cap W_2^2(0, T; \hat{\mathscr{H}})$. The result for problem P_ε follows from Proposition 2. □

4 Proof of Theorem 1

We begin by estimating the penalized solutions $\hat{\mathbf{u}}_\varepsilon$, which enable us to pass to the limit, and hence to obtain a solution to problem $\hat{P}$.

4.1 Estimates of Solution to Problem $\hat{P}_\varepsilon$

Let $\hat{\mathbf{w}} = (0, 0)$ in (38) and integrate over $[0, t]$, $t \in (0, T)$. Then we have

$$\begin{aligned} & \int_0^t (\partial_t^2 \hat{\mathbf{u}}_\varepsilon, \partial_t \hat{\mathbf{u}}_\varepsilon)_{\hat{\mathscr{H}}} \, \mathrm{d}t + \int_0^t \hat{a}(\partial_t \hat{\mathbf{u}}_\varepsilon, \partial_t \hat{\mathbf{u}}_\varepsilon) \, \mathrm{d}t + \int_0^t \hat{b}(\hat{\mathbf{u}}_\varepsilon, \partial_t \hat{\mathbf{u}}_\varepsilon) \, \mathrm{d}t \\ & \quad + \int_0^t \hat{\Phi}_\varepsilon(\hat{\mathbf{u}}_\varepsilon, \partial_t \hat{\mathbf{u}}_\varepsilon) \, \mathrm{d}t \leq \int_0^t \langle \hat{\mathbf{L}}, \partial_t \hat{\mathbf{u}}_\varepsilon \rangle_{\hat{\mathscr{V}}} \, \mathrm{d}t, \end{aligned}$$

and hence for almost every $t \in (0, T)$

$$
\begin{aligned}
&\frac{1}{2}\left|\partial_t \hat{\mathbf{u}}_\varepsilon(t)\right|^2_{\hat{\mathscr{H}}} + \int_0^t \hat{a}(\partial_t \hat{\mathbf{u}}_\varepsilon, \partial_t \hat{\mathbf{u}}_\varepsilon)\,\mathrm{d}t + \frac{1}{2}\hat{b}(\hat{\mathbf{u}}_\varepsilon(t), \hat{\mathbf{u}}_\varepsilon(t)) \\
&+\frac{1}{2\varepsilon}\left\|\left([\hat{u}_{\varepsilon N}(t)] - g\right)_+\right\|^2_{L_2(\Xi)} \le \int_0^t \langle \hat{\mathbf{L}}, \partial_t \hat{\mathbf{u}}_\varepsilon\rangle_{\hat{\mathscr{V}}}\,\mathrm{d}t + \frac{1}{2}|\hat{\mathbf{v}}_0|^2_{\hat{\mathscr{H}}} + \frac{1}{2}\hat{b}(\hat{\mathbf{u}}_0, \hat{\mathbf{u}}_0).
\end{aligned}
$$

By virtue of (39) and Young's inequality, there exists a positive constant M, independent of ε, such that the following estimates hold for any $\varepsilon > 0$:

$$
\begin{cases}
\left|\hat{\mathbf{v}}_\varepsilon(t)\right|_{\hat{\mathscr{H}}} \le M \;\; a.e.\; t \in (0, T), & \displaystyle\int_0^T \|\hat{\mathbf{v}}_\varepsilon\|^2_{\hat{\mathscr{V}}}\,\mathrm{d}t \le M, \\
\left\|\hat{\mathbf{u}}_\varepsilon(t)\right\|_{\hat{\mathscr{V}}} \le M, & \left\|\left([\hat{u}_{\varepsilon N}(t)] - g\right)_+\right\|_{L_2(\Xi)} \le M\sqrt{\varepsilon}\;\; \text{a.e. } t \in (0, T).
\end{cases} \tag{42}
$$

From (38) we can derive for any $\hat{\phi} = (\phi^+, \phi^-) \in L_2(0, T; H^1_0(\Omega^+) \times H^1_0(\Omega^-))$

$$
\int_0^T (\partial_t^2 \hat{\mathbf{u}}_\varepsilon, \hat{\phi})_{\hat{\mathscr{H}}}\,\mathrm{d}t + \int_0^T \hat{a}(\partial_t \hat{\mathbf{u}}_\varepsilon, \hat{\phi})\,\mathrm{d}t + \int_0^T \hat{b}(\hat{\mathbf{u}}_\varepsilon, \hat{\phi})\,\mathrm{d}t = \int_0^T (\hat{\mathbf{f}}, \hat{\phi})_{\hat{\mathscr{H}}}\,\mathrm{d}t.
$$

This yields that $\partial_t^2 \hat{\mathbf{u}}_\varepsilon$ is bounded in $L_2(0, T; H^{-1})$ by a constant independent of ε.

For any $\check{\mathbf{w}} \in L_\infty(0, T; \hat{\mathscr{V}}) \cap W^1_2(0, T; \hat{\mathscr{H}})$, $\check{\mathbf{w}}(t) \in \hat{\mathscr{K}}$ a.e. $t \in (0, T)$, we take in (38) $\hat{\mathbf{w}} = \hat{\mathbf{v}}_\varepsilon + \check{\mathbf{w}} - \hat{\mathbf{u}}_\varepsilon$. Then, integrating it over $(0, T)$, we have by virtue of the integration by parts and a monotonicity property of $\hat{\Phi}_\varepsilon$

$$
\begin{aligned}
&(\hat{\mathbf{v}}_\varepsilon(T), \check{\mathbf{w}}(T) - \hat{\mathbf{u}}_\varepsilon(T))_{\hat{\mathscr{H}}} - (\hat{\mathbf{v}}_0, \check{\mathbf{w}}(0) - \hat{\mathbf{u}}_0)_{\hat{\mathscr{H}}} \\
&\quad+ \int_0^T \left\{-(\hat{\mathbf{v}}_\varepsilon, \partial_t \check{\mathbf{w}})_{\hat{\mathscr{H}}} + \hat{a}(\hat{\mathbf{v}}_\varepsilon, \check{\mathbf{w}}) + \hat{b}(\hat{\mathbf{u}}_\varepsilon, \check{\mathbf{w}}) + \hat{J}(\hat{\mathbf{u}}_\varepsilon, \hat{\mathbf{v}}_\varepsilon, \hat{\mathbf{v}}_\varepsilon + \check{\mathbf{w}} - \hat{\mathbf{u}}_\varepsilon)\right\} dt \\
&\quad\ge \int_0^T \langle \hat{\mathbf{L}}, \check{\mathbf{w}} - \hat{\mathbf{u}}_\varepsilon\rangle_{\hat{\mathscr{V}}}\,\mathrm{d}t \\
&\qquad+ \int_0^T \left\{-|\hat{\mathbf{v}}_\varepsilon|^2_{\hat{\mathscr{H}}} + \hat{a}(\hat{\mathbf{v}}_\varepsilon, \hat{\mathbf{u}}_\varepsilon) + \hat{b}(\hat{\mathbf{u}}_\varepsilon, \hat{\mathbf{u}}_\varepsilon) + \hat{J}(\hat{\mathbf{u}}_\varepsilon, \hat{\mathbf{v}}_\varepsilon, \hat{\mathbf{v}}_\varepsilon)\right\}\,\mathrm{d}t.
\end{aligned} \tag{43}
$$

4.2 Passing to the Limit: Solvability of Problem $\hat{P}$

From the estimates in (42) and the previous estimate of $\partial_t^2\hat{\mathbf{u}}_\varepsilon$, it follows that there exists $\hat{\mathbf{u}} = (\mathbf{u}^+, \mathbf{u}^-)$ such that, up to a subsequence,

$$\begin{cases} \hat{\mathbf{u}}_\varepsilon \overset{w^*}{\rightharpoonup} \hat{\mathbf{u}} \text{ in } L_\infty(0,T;\hat{\mathscr{V}}), \\ \partial_t\hat{\mathbf{u}}_\varepsilon \overset{w}{\rightharpoonup} \partial_t\hat{\mathbf{u}} \text{ in } L_2(0,T;\hat{\mathscr{V}}), \quad \partial_t\hat{\mathbf{u}}_\varepsilon \overset{w^*}{\rightharpoonup} \partial_t\hat{\mathbf{u}} \text{ in } L_\infty(0,T;\hat{\mathscr{H}}), \\ \partial_t^2\hat{\mathbf{u}}_\varepsilon \overset{w}{\rightharpoonup} \partial_t^2\hat{\mathbf{u}} \text{ in } L_2(0,T;H^{-1}) \quad \text{as } \varepsilon \to 0. \end{cases} \tag{44}$$

Owing to (44) one can easily pass to the limit in the linear terms of (43). For the nonlinear terms it is necessary to rely on Theorem 2. Since $\partial\Omega$ is Lipschitz continuous, the embeddings from $\mathscr{V}^\pm$ into $\mathscr{H}^\pm$, from $\mathscr{V}^\pm$ into $H^{1/2}(\Omega^\pm)$ and from $\mathscr{H}^\pm$ into $H^{-1/2}(\Omega^\pm)$ are compact. Accordingly, we may apply Theorem 2 with

$$\begin{aligned} &\mathscr{F} = \{\partial_t\hat{\mathbf{u}}_\varepsilon\}_{\varepsilon>0}, \quad X = \hat{\mathscr{V}}, \quad U = \hat{\mathscr{H}}, \quad Y = H^{-1}, \quad p = 2, \\ &\mathscr{F} = \{\hat{\mathbf{u}}_\varepsilon\}_{\varepsilon>0}, \quad X = \hat{\mathscr{V}}, \quad U = H^{1/2}, \quad Y = \hat{\mathscr{H}}, \quad r = 2, \\ &\mathscr{F} = \{\partial_t\hat{\mathbf{u}}_\varepsilon\}_{\varepsilon>0}, \quad X = \hat{\mathscr{H}}, \quad U = H^{-1/2}, \quad Y = H^{-1}, \quad p = 2, \end{aligned}$$

so that, up to a subsequence, we get as $\varepsilon \to 0$

$$\begin{cases} \partial_t\hat{\mathbf{u}}_\varepsilon \to \partial_t\hat{\mathbf{u}} & \text{in } L_2(0,T;\hat{\mathscr{H}}), \\ \hat{\mathbf{u}}_\varepsilon \to \hat{\mathbf{u}} & \text{in } C([0,T];H^{1/2}), \\ \partial_t\hat{\mathbf{u}}_\varepsilon \to \partial_t\hat{\mathbf{u}} & \text{in } C([0,T];H^{-1/2}), \end{cases} \tag{45}$$

and hence $(\hat{\mathbf{v}}_\varepsilon(T), \check{\mathbf{w}}(T) - \hat{\mathbf{u}}_\varepsilon(T))_{\hat{\mathscr{H}}} \to \langle\hat{\mathbf{v}}(T), \check{\mathbf{w}}(T) - \hat{\mathbf{u}}(T)\rangle_{-1/2,1/2}$ as $\varepsilon \to 0$. The functional $\mathbf{w} \mapsto \int_0^T \hat{b}(\mathbf{w},\mathbf{w})\,dt$ is convex and continuous on $L_2(0,T;\hat{\mathscr{V}})$, so that it is sequentially weakly lower semi-continuous. Therefore,

$$\liminf_{\varepsilon\to 0} \int_0^T \hat{b}(\hat{\mathbf{u}}_\varepsilon, \hat{\mathbf{u}}_\varepsilon)\,dt \ge \int_0^T \hat{b}(\hat{\mathbf{u}}, \hat{\mathbf{u}})\,dt. \tag{46}$$

The fact $W_2^1(0,T;\hat{\mathscr{V}}) \subset C([0,T];\hat{\mathscr{V}})$ implies that, for all $t \in [0,T]$, $\{\hat{\mathbf{u}}_\varepsilon(t)\}_{\varepsilon>0}$ is bounded in $\hat{\mathscr{V}}$ by a constant independent of ε. Thus, by a diagonal process we can extract a subsequence (still denoted by $\{\hat{\mathbf{u}}_\varepsilon\}_{\varepsilon>0}$) such that

$$\hat{\mathbf{u}}_\varepsilon(t) \overset{w}{\rightharpoonup} \hat{\mathbf{u}} \quad \text{in } \hat{\mathscr{V}} \quad (\varepsilon \to 0), \quad \forall t \in [0,T]. \tag{47}$$

Since $\mathbf{w} \mapsto \hat{a}(\mathbf{w},\mathbf{w})$ is sequentially weakly lower semi-continuous on $\hat{\mathscr{V}}$, we obtain by (47)

$$\liminf_{\varepsilon\to 0}\int_0^T \hat{a}(\hat{\mathbf{v}}_\varepsilon, \hat{\mathbf{u}}_\varepsilon)\,\mathrm{d}t = \liminf_{\varepsilon\to 0}\frac{1}{2}\left(\hat{a}(\hat{\mathbf{u}}_\varepsilon(T), \hat{\mathbf{u}}_\varepsilon(T)) - \hat{a}(\hat{\mathbf{u}}_0, \hat{\mathbf{u}}_0)\right)$$

$$\geq \frac{1}{2}\left(\hat{a}(\hat{\mathbf{u}}(T), \hat{\mathbf{u}}(T)) - \hat{a}(\hat{\mathbf{u}}_0, \hat{\mathbf{u}}_0)\right) = \int_0^T \hat{a}(\hat{\mathbf{v}}, \hat{\mathbf{u}})\,\mathrm{d}t. \tag{48}$$

Since the embedding from $H^{1/2}(\Xi)$ into $L_2(\Xi)$ is compact, it follows from (47)

$$[\hat{u}_{\varepsilon N}](t) \to [\hat{u}_N](t) \;\text{ in } L_2(\Xi) \;\; (\varepsilon \to 0), \;\; \forall t \in [0, T].$$

Accordingly, $(42)_4$ leads to

$$\lim_{\varepsilon\to 0} \|([\hat{u}_{\varepsilon N}](t) - g)_+\|_{L_2(\Xi)} = \|([\hat{u}_N](t) - g)_+\|_{L_2(\Xi)} = 0, \;\; \forall t \in [0, T].$$

Thus, $[\hat{u}_N](t) \leq g$ holds almost everywhere on Ξ and for any $t \in [0, T]$, which concludes $\hat{\mathbf{u}}(t) \in \hat{\mathscr{K}}$ for any $t \in [0, T]$.

For the friction term, we again apply Theorem 2 with $\mathscr{F} = \{\hat{\mathbf{v}}_\varepsilon = \partial_t \hat{\mathbf{u}}_\varepsilon\}_{\varepsilon>0}$, $X = \hat{\mathscr{V}}$, $U = H^{1-\delta}$ ($\delta \in (0, 1/2)$), $Y = H^{-1}$, $p = 2$, so that we obtain

$$\hat{\mathbf{u}}_\varepsilon \to \hat{\mathbf{u}} \;\text{ in } W_2^1(0, T; H^{1-\delta}) \;\; (\varepsilon \to 0).$$

Since the mapping $\mathbf{w} \mapsto [\mathbf{w}_\tau] : H^{1-\delta} \to L_2(\Xi)$ ($\delta \in (0, 1/2)$) is compact,

$$[\hat{\mathbf{u}}_{\varepsilon\tau}] \to [\hat{\mathbf{u}}_\tau] \;\text{ in } W_2^1(0, T; L_2(\Xi)) \;\; (\varepsilon \to 0). \tag{49}$$

One can pass to the lower limit in (43) with the help of (12), $(42)_2$, and (49), so that (19) is established. In consequence, we have solved problem $\hat{P}$.

Finally, we define $\mathbf{u}$ by $\mathbf{u} = \mathbf{u}^+$ on Ω^+ and $\mathbf{u} = \mathbf{u}^-$ on Ω^-, which satisfies (15). In conclusion, the proof of Theorem 1 is completed by applying Proposition 1.

References

1. Boieri, P., Gastaldi, F., Kinderlehrer, D.: Existence, uniqueness, and regularity results for the two-body contact problem. Appl. Math. Optim. **15**, 251–277 (1987)
2. Brézis, H.: Problèmes unilatéraux. J. Math. Pures et Appl. **51**, 1–168 (1972)
3. Chernikh, K.F.: Introduction to the Physical and Geometrical Nonlinear Theory of Fracture. Nauka (1996). (Russian)
4. Cocou, M.: Existence of solutions of Signorini problems with friction. Int. J. Eng. Sci. **22**, 567–575 (1984)
5. Cocou, M.: Existence of solutions of a dynamic Signorini's problem with nonlocal friction in viscoelasticity. Z. Angew. Math. Phys. **53**, 1099–1109 (2002)
6. Cocou, M.: A class of dynamic contact problems with Coulomb friction in viscoelasticity. Nonl. Analysis: Real World Appl. **22**, 508–519 (2015)

7. Cocou, M., Ricaud, J.M.: Analysis of a class of implicit evolution inequalities associated to viscoelastic dynamic contact problems with friction. Int. J. Eng. Sci. **38**, 1535–1552 (2000)
8. Cocou, M., Scarella, G.: Existence of a solution to a dynamic unilateral contact problem for a cracked viscoelastic body. C.R. Math. Acad. Sci. Paris **338** (4), 341–346 (2004)
9. Cocou, M., Scarella, G.: Analysis of a dynamic unilateral contact problem for a cracked viscoelastic body. Z. Angew. Math. Phys. **57**, 523–546 (2006)
10. Cocou, M., Schryve, G., Raous, M.: A dynamic unilateral contact problem with adhesion and friction in viscoelasticity. Z. Angew. Math. Phys. **61**, 721–743 (2010)
11. Duvaut, G.: G. Equilibre d'un solide elastique avec contact unilateral et frottement de Coulomb. C. R. Acad. Sci. Paris, Ser. A **290**, 263–265 (1980)
12. Duvaut, G., Lions, J.L.: Les inéquations en mécanique et en physique. Dunod, Paris (1972)
13. Eck, C., Jarušek, J., Krbec, M.: Unilateral Contact Problems—Variational Methods and Existence Theorems. Chapman and Hall/CRC (2005)
14. Fichera, G.: Boundary value problems of elasticity with unilateral constraints. In: Handbuch der Physik, vol. VIa/2, pp. 391–424 (1972)
15. Freund, L.B.: Dynamic Fracture Mechanics. Cambridge University Press, Cambridge (1998)
16. Glowinski, R., Lions, J.L., Trémolières, R.: Analyse Numérique des Inéquations Variationnelles. Dunod, Paris (1976)
17. Guz, A.N., Zozulya, V.V.: Elastodynamic unilateral contact problems with friction for bodies with cracks. Int. Appl. Mekh. **38**, 895–932 (2002)
18. Haupt, P.: Continuum Mechanics and Theory of Materials. Springer, Berlin (2000)
19. Jarušek, J.: Dynamic contact problems with given friction for viscoelastic bodies. Czech. Math. J. **46**, 475–487 (1996)
20. Kanamori, H.: Mechanics of eqrthquakes. Ann. Rev. Earth Planet. Sci. **22**, 207–237 (1994)
21. Kanamori, H., Brodsky, E.E.: The physics of eqrthquakes. Rep. Prog. Phys. **67**, 1429–1496 (2004)
22. Kuttler, K.L., Shillor, M.: Dynamic bilateral contact with discontinuous friction coefficient. Nonlinear Anal. **45**, 309–327 (2001)
23. Kuttler, K.L., Shillor, M.: Dynamic contact with Signorini's condition and slip rate dependent friction. Electron. J. Differ. Equ. **2004**(83), 1–21 (2004)
24. Landau, L.D., Lifshits, E.M.: Theory of Elasticity. Nauka (1987). (Russian)
25. Leitman, M.J., Fisher, G.M.: The linear theory of viscoelasticity. In: Handbuch der Physik, vol. VIa/3, pp. 1–123 (1973)
26. Martins, J.A.C., Oden, J.T.: Existence and uniqueness results for dynamic contact problems with nonlinear normal and friction, interface laws. Nonlinear Anal. **11**, 407–428 (1987)
27. Persson, B.N.J.: Sliding Friction Physical Principles and Applications. Springer, Berlin (2000)
28. Ricaud, J.M., Pratt, E.: Analysis of a time discretisation for an implicit variational inequality modelling dynamic contact problems with friction. Math. Meth. Appl. Sci. **24**, 491–511 (2001)
29. Scholz, C.H.: The Mechanics of Earthquakes and Faulting, 2nd edn. Cambridge, UP (2002)
30. Simon, J.: Compact sets in the space $L_p(0, T; B)$. Ann. Mat. Pure Appl. **146**, 65–96 (1987)

Some Recent Results on Regularity of the Crack-tip/Crack-front of Mumford–Shah Minimizers

Hayk Mikayelyan

Abstract In this short note, we would like to present some recent results about the regularity of the crack at the crack-tip/crack-front for the minimizers of the Mumford–Shah functional in $\mathbf{R}^2$ and $\mathbf{R}^3$. The proofs can be found in [6, 25, 27].

1 Introduction

The Mumford–Shah functional

$$J(u,\Gamma) := \int\limits_{\Omega\setminus\Gamma} |\nabla u|^2 \; + \alpha(u-h)^2\,dx + \beta\mathscr{H}^{N-1}(\Gamma),$$

with $g \in L^\infty(\Omega)$ and $u \in H^1(\Omega \setminus \Gamma)$, has been introduced by David Mumford and Jayant Shah in [28]. It was used in image processing to find a good approximation u for the raw image data given by the function h. The discontinuity set Γ then locates the edges of the objects. The idea of competing bulk and surface energies, however, goes back to Alan Arnold Griffith (see [16]), whose theory of brittle fracture is based on the balance between gain in surface energy and strai n energy release. Francfort and Marigo used this concept to develop now classical variational model for crack propagation in [15].

In this article, we consider the simplified model with $\alpha = 0$, $\beta = 1$,

$$J(u,\Gamma) := \int\limits_{\Omega\setminus\Gamma} |\nabla u|^2 dx + \mathscr{H}^{N-1}(\Gamma). \tag{1}$$

The absence of the second term ($\alpha = 0$) will imply constant minimizers and to avoid this we need to impose boundary conditions on $\partial\Omega$. Let us discuss local minimizers

H. Mikayelyan (✉)
University of Nottingham Ningbo, Ningbo 315100, China
e-mail: Hayk.Mikayelyan@nottingham.edu.cn

H. Itou et al. (eds.), *Mathematical Analysis of Continuum Mechanics and Industrial Applications III*, Mathematics for Industry 34,
https://doi.org/10.1007/978-981-15-6062-0_2

defined as follows. For an open set $\Omega \subset \mathbf{R}^N$, let (u, Γ) be a couple such that $\Gamma \subset \Omega$ is closed and $u \in H^1(B \setminus \Gamma)$, for any ball $B \subset \Omega$. Let us consider the local Mumford–Shah energy in a ball B

$$J(u, \Gamma, B) := \int\limits_{B\setminus\Gamma} |\nabla u|^2\, dx + \mathscr{H}^{N-1}(\Gamma \cap B),$$

and local minimizers (u, Γ) in Ω, i.e., such that

$$J(u, \Gamma, B) \leq J(u', \Gamma', B)$$

for any $B \subset \Omega$ and for any competitor (u', Γ') which satisfies $u = u'$ in $\Omega \setminus B$ and $\Gamma = \Gamma'$ in $\Omega \setminus B$. In that case, we will simply say that (u, Γ) is a Mumford–Shah minimizer.

The Euler–Lagrange equation associated to the Mumford–Shah functional can be derived in terms of domain variation, which is given by any $\eta \in C_c^1(\Omega)^N$ and implies the equation (see [3, Theorem 7.35])

$$\int\limits_{\Omega\setminus\Gamma} |\nabla u|^2 \operatorname{div} \eta - 2\langle \nabla u, \nabla u \cdot \nabla \eta\rangle\, dx + \int\limits_{\Gamma} \operatorname{div}^{\Gamma} \eta\, d\mathscr{H}^{N-1} = 0. \tag{2}$$

A couple (u, Γ) satisfying the equation (2) for all $\eta \in C_c^1(\Omega)^N$ will be called stationary in Ω.

Ennio De Georgi, Michele Carriero, and Antonio Leaci in their pioneering work [12] have shown the existence of the minimizers of Mumford–Shah functional and Luigi Ambrosio, Alexis Bonnet, Guy David, Nicola Fusco, and Diego Pallara (see [1, 2, 4, 8–10]) developed the regularity theory.

Two famous monographs [3, 11] as well as the survey paper of Antoine Lemenant [24] contain most of the known results. It is also worth to mention some recent publications, such as [13, 14, 22].

The most interesting example of a Mumford–Shah minimizer in the plane is the crack-tip function, which is known to be the only non-constant element among the global minimizers (see [8]). This function plays a key role in brittle fracture theory and is defined as follows:

$$\Gamma_0 := \{(x, 0) \mid x \leq 0\}, \tag{3}$$

and

$$u_0(r, \theta) := \sqrt{\frac{2r}{\pi}} \sin(\theta/2) \quad r > 0, \quad \theta \in (-\pi, \pi). \tag{4}$$

It has been proven by Bonnet and David in [9] that the couple (u_0, Γ_0) is a Mumford–Shah minimizer in the plane.

Despite being the only global minimizer with non-constant elements, the crack-tip function is of particular interest because it is the only known singularity with the surface energy and the bulk energy in a ball B_r scaling of the same order as $r \to 0$. This makes impossible the use of the methods developed for minimal surface equation, as in case of other types of singularities, where the surface term dominates over the bulk term.

In the next three sections, we will introduce without proofs the results recently obtained in [6, 25, 27].

2 Regularity Up to the Crack-Tip in 2D

The main motivation in studying the regularity of the Mumford–Shah minimizers at the crack-tip comes from the variational model of Francfort and Marigo. Without going into detail, let us observe that if we assume that the crack Γ is a C^1-curve with a crack-tip in the origin, then the Euler–Lagrange equation (2) will imply the following conditions:

$$\Delta u = 0 \quad \text{in} \quad \Omega \setminus \Gamma, \tag{5}$$

$$\partial_\nu u^\pm \Big|_\Gamma = 0 \quad \text{on} \quad \Gamma, \tag{6}$$

$$H_\Gamma = \left[|\nabla u^+|^2 - |\nabla u^-|^2 \right] \Big|_\Gamma \quad \text{on} \quad \Gamma, \tag{7}$$

$$u(r, \phi) = u(0, 0) + \sqrt{\frac{2}{\pi}} r^{1/2} \sin(\phi/2) + o(r^{1/2}), \tag{8}$$

where H_Γ is the curvature of Γ, u^+ and u^- are the values of u on different sides of Γ, and (r, ϕ) is a certain choice of polar coordinates in $\mathbf{R}^2$ (details can be found in [3]). Using simple heuristic arguments, one can see that when approaching the crack-tip the curvature can explode. This is not a very natural thing to happen for a minimizer of the functional which involves a perimeter term. Indeed, from Theorem 1 it follows that the curvature must vanish at the crack-tip. It is a matter of ongoing research to exploit this fact in deriving better estimates for the brittle fracture model of Francfort and Marigo.

In [6], John Andersson and the author introduce the following definition.

Definition 1 We say that (u, Γ) is $\varepsilon-$close to a crack-tip if the following holds:

1. (u, Γ) is a minimizer of (1) in $B_1 = B_1(0) (= \Omega)$ with some specified boundary conditions.
2. Γ consists of a connected rectifiable curve that starts at the origin and connects the origin to ∂B_1, i.e., there exists a Lipschitz mapping $\tau : [0, 1] \mapsto \overline{B_1}$ such that $\tau(0) = 0$, $\tau(1) \in \partial B_1$ and $\Gamma = \tau([0, 1])$.
3. For some $\lambda \in \mathbf{R}$,

$$\left(\int\limits_{B_1 \setminus \Gamma} \left| \nabla \left(u - \lambda \sqrt{\frac{2}{\pi}} r^{1/2} \sin(\phi/2) \right) \right|^2 \mathrm{d}x \right)^{\frac{1}{2}} \leq \varepsilon,$$

where (r, ϕ) are the standard polar coordinates of $\mathbf{R}^2$.

4. And $u(0, 0) = 0$.

The proof of the following theorem can be found in [6].

Theorem 1 *There exists an $\varepsilon_0 > 0$ such that if (u, Γ) is ε-close to a crack-tip solution for some $\varepsilon \leq \varepsilon_0$, then Γ is $C^{2,\alpha}$ at the crack-tip for every $\alpha < \alpha_2 - 3/2$, $\alpha_2 \approx 1.889$.*

This in the sense that the tangent at the crack-tip is a well-defined line, which we may assume (after rotating the coordinate system) to be $\{(x, 0);\ x \in \mathbf{R}\}$, and there exists a constant C_α such that

$$\Gamma \subset \left\{ (x, y);\ |y| < C_\alpha \varepsilon |x|^{2+\alpha},\ x < 0 \right\}.$$

Here C_α may depend on α but not on $\varepsilon < \varepsilon_0$.

The proof is based on linearization technique developed in recent years by John Andersson to treat certain free boundary problems. The authors adapt this method to the free discontinuity context to calculate the asymptotic expansion of the function u at the crack-tip. In particularly, it is proven that a blow-up sequence u_j, $j \geq 1$, can be expressed as follows:

$$u_j = \frac{2}{\pi} r^{1/2} \sin \phi/2 + \varepsilon_j v(r, \phi) + R_j,$$

where $R_j << \varepsilon_j$, and

$$v(r, \phi) = \sum_{k=2}^{\infty} a_k r^{\alpha_k} \cos(\alpha_k \phi) + \sum_{k=2}^{\infty} b_k r^{k-1/2} \sin((k - 1/2)\phi). \tag{9}$$

Here the powers $k - 1 < \alpha_k < k$, for $k = 2, 3, 4 \ldots$ satisfy the relation

$$\tan(\alpha\pi) = -\frac{2}{\pi} \frac{\alpha}{\alpha^2 - \frac{1}{4}}. \tag{10}$$

Theorem 1 in particularly means that the Euler–Lagrange conditions (5)–(8) were not complete, and there was a missing Euler–Lagrange condition. We refer the interested reader to [6] for further details.

3 Fine Numerical Analysis of the Crack-Tip Position

When Theorem 1 was still a conjecture, the author and Zhilin Li analyzed numerically the curvature of the crack near the crack-tip. The main obstacle for this is the degeneracy of the function u near the crack-tip, which makes a reliable verification of the condition (7) impossible. To address this problem, the conformal mapping was used.

Let (u, Γ) be the minimizer of the Mumford–Shah functional in the ball $B_1 = B_1(0)$, and let the boundary values of u on ∂B_1 be a perturbation of those of the global minimizer $\sqrt{\frac{2}{\pi}} r^{1/2} \sin(\phi/2)$. Further let us assume that Γ is a curve connecting the boundary with a point (= the crack-tip) in the interior of B_1. In [27], we introduce a numerical algorithm, which determines Γ with high accuracy.

First, let us observe that if we place the origin in the crack-tip and apply the conformal mapping $\sqrt{z}$ with the branch-cut Γ we will obtain a new function $\tilde{u}(\tilde{x}, \tilde{y})$ defined in the domain Ω (see Figs. 1 and 2). The boundary of Ω consists of two parts: $(\partial\Omega)_D$ which is the image of ∂B_1, and where the function $\tilde{u}$ will satisfy Dirichlet conditions, and $(\partial\Omega)_N$, symmetric with respect to the origin. which is the image of the both faces of Γ. Since Laplace operator is invariant under conformal mappings, the function $\tilde{u}$ will be harmonic (Eq. (11)), and since angles are preserved the function $\tilde{u}$ will satisfy the Neumann boundary conditions on $(\partial\Omega)_N$ (equation (12)). Further properties of $\tilde{u}$ inherited from (7) and (8) are (13)–(14).

$$\Delta\tilde{u} = 0 \quad \text{in} \quad \Omega \tag{11}$$

$$\partial_\nu \tilde{u}\Big|_{\partial\Omega} = 0 \quad \text{on} \quad (\partial\Omega)_N \tag{12}$$

$$(\tilde{x}^2 + \tilde{y}^2)^{\frac{1}{2}} H_{(\partial\Omega)_N}(\tilde{x}, \tilde{y}) + \frac{\tilde{x}\tilde{y}' - \tilde{x}'\tilde{y}}{(\tilde{x}^2 + \tilde{y}^2)^{\frac{1}{2}} (\tilde{x}'^2 + \tilde{y}'^2)^{\frac{1}{2}}} = \frac{|\partial_\tau \tilde{u}|^2 (\tilde{x}, \tilde{y}) - |\partial_\tau \tilde{u}|^2 (-\tilde{x}, -\tilde{y})}{2} \tag{13}$$

$$|\nabla\tilde{u}(0, 0)| = \sqrt{\frac{2}{\pi}}, \tag{14}$$

where in the equation (13) $(\tilde{x}, \tilde{y})$ parametrize the boundary $(\partial\Omega)_N$ and $H_{(\partial\Omega)_N}$ is its curvature. The solution $\tilde{u}$ of the problem (11)–(14) is not degenerate, and its gradient is bounded.

To solve the original problem, we first solve another problem: minimize the Mumford–Shah functional among all pairs (u, Γ), where the curve Γ connects the boundary with a **given** point in $B_{1/4}$. In other words, we prescribe location of the crack-tip. After solving this problem, we measure the Mumford–Shah energy and at the end we pick the crack-tip which corresponds to least Mumford–Shah energy. The construction of the solutions and the measurements are done with high accuracy as follows:

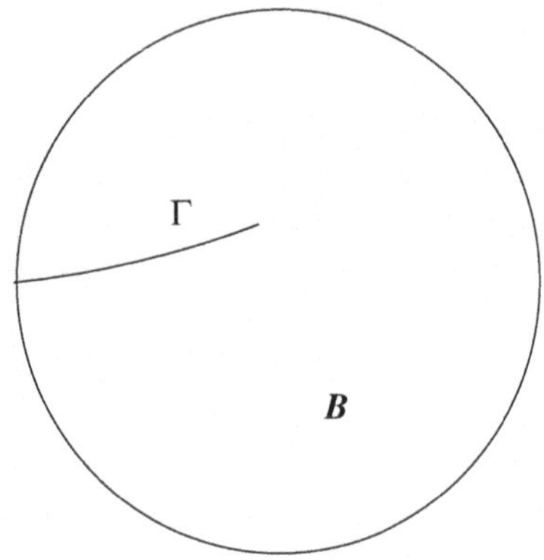

Fig. 1 A diagram of a crack Γ. ©Eur. Mathematical Society, see [27]

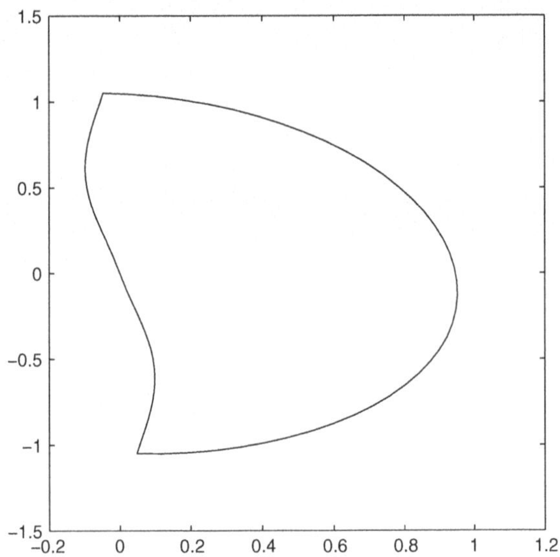

Fig. 2 Domain Ω after transformation $\sqrt{z}$. ©European Mathematical Society, see [27]

Step 1. (Problem setting)

Prescribe a crack-tip by choosing a point in $B_{1/4}$. Shift the coordinate system such that the origin is in that point. Connect the point on the boundary, where the boundary data has discontinuity, with the origin using a certain parabola Γ_1 (initial guess for the crack).

Step 2. (Finding the solution)

Apply the conformal mapping $\sqrt{z}$ with branch-cut Γ to obtain the domain Ω_1. Solve the overdetermined problem (11)–(13) using the following iteration:
(a) Solve mixed Dirichlet–Neumann boundary value problem (11) and (12) in Ω_1 to get the solution $\tilde{u}_1$.

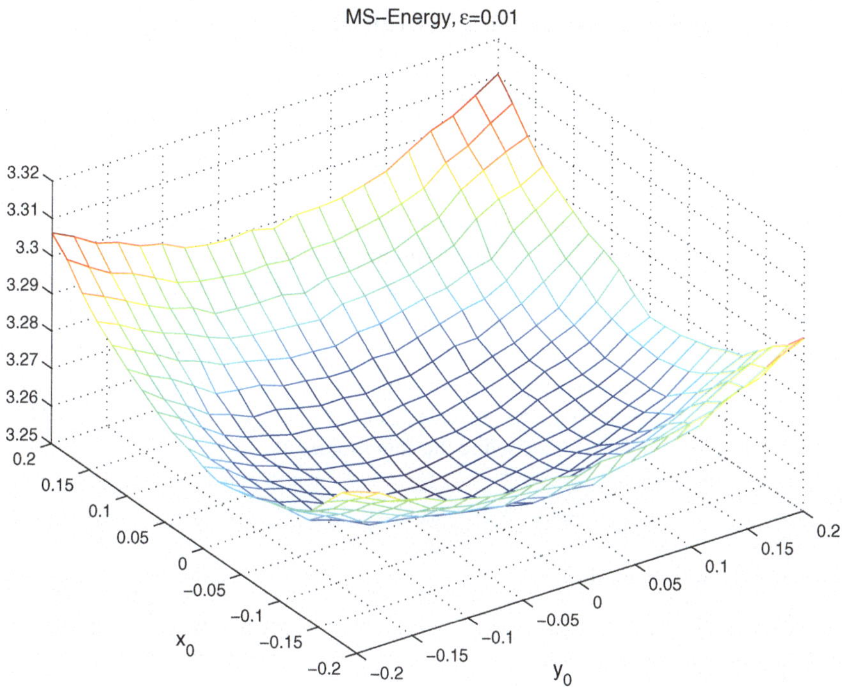

Fig. 3 Plot of the Mumford–Shah energy as function of the prescribed crack-tip in $B_{1/5}$. ©European Mathematical Society, see [27]

(b) The condition (13) is not necessarily satisfied by our initial guess for $(\partial\Omega_1)_N$, and the solution $\tilde{u}_1$ obtained in (a). We adjust the boundary $(\partial\Omega_1)_N$ by fixing the data of $\tilde{u}_1$, and solving (13) as a second-order ODE with respect to $(\tilde{x}, \tilde{y})$. As a result, we obtain a new domain Ω_2.
(c) Go to (a) and continue the iteration until (13) is satisfied (small error term).

Step 3. (Measurements)

Compute $\int_{B_1\setminus\Gamma} |\nabla u|^2 dxdy = \int_{\Omega} |\nabla\tilde{u}|^2 d\tilde{x}d\tilde{y}$ in the set Ω, where $\tilde{u}$ is not degenerate. Obtain the crack Γ by applying the inverse conformal mapping z^2. Compute the length of Γ and the full Mumford–Shah energy. Determine the optimal crack-tip by finding the one with the minimal energy (see Fig. 3). Compute the curvature of Γ in the crack-tip using nonlinear least square fitting with appropriate functions to approximate the graph of $(\partial\Omega)_N$ near the origin.

Using this, we have verified numerically the conjecture later proven in [6]. We refer the interested reader to [27] for further details.

4 Stationarity of the Crack-Front in 3D

Theorem 1 established a new previously unknown Euler–Lagrange condition for the minimizers of the Mumford–Shah functional in 2D. This obviously means that there is a missing condition in 3D as well (possible several conditions).

Let us assume that (u, P) is a minimizer of the Mumford–Shah functional in 3D, where P is a two-dimensional smooth surface with a crack-front γ, given by a smooth curve. The author believes that the key to finding new Euler–Lagrange condition(s) near the crack-front γ lies in understanding the dependence of the minimizing function $u(x)$ on $x \in \gamma$. Some basic research in this direction has been initiated with Antoine Lemenant in [25].

Let us extend the crack-tip global solution (u_0, Γ_0) defined in (3) and (4) into 3D and define the pair (u_0, P_0) as $u_0(x, y, z) = u_0(x, y)$, and $P_0 = \Gamma_0 \times \mathbf{R} = \{x, 0, z \mid x \leq 0\}$. Further, let us define the function u_δ as

$$u_\delta(x, y, z) = u_0(x, y, z) + \delta z.$$

We prove the following theorem.

Theorem 2 *The pair* (u_δ, P_0) *is stationary in* $\mathbf{R}^3$ *for any* $\delta \in \mathbf{R}$, *but is minimizing in* $\mathbf{R}^3$ *if and only if* $\delta = 0$.

In [8], topological minimizers of Mumford–Shah functional are introduced. For example, it is easy to see that in 2D the pair (u_L, L), where $u_L(x, y) = \mathrm{sgn}(y)$, and $L = \{(x, 0) \mid x \in \mathbf{R}\}$, is stationary. On the other side, it is not a minimizer if one is allowed to make holes in the line L. However, if one is not allowed to change the topology of $\mathbf{R}^2 \setminus L$, the pair is a minimizer, or topological minimizer. To the best of our knowledge, (u_δ, P_0) is the first known example of a stationary solution, which is not a topological minimizer (see [25, Remark 6.1]).

Further, we consider the following Mumford–Shah minimization problem in the cylinder $\mathscr{C} := \{(x, y, z) \mid x^2 + y^2 < 1 \text{and} - 1 < z < 1\}$:

$$\min\left\{ \int_{\mathscr{C}\setminus P} |\nabla u|^2\, dx + \mathscr{H}^2(P)\ ;\ u = u_\delta \text{on} \partial\mathscr{C} \right\}. \tag{15}$$

We prove the following proposition.

Proposition 1 *If* $\delta > \sqrt{3 - \frac{2}{\pi}}$ *then* u_δ *is not a minimizer of the problem (15).*

It is very natural to ask the following.

Proposition 2 *Is* u_δ *a solution of Problem (15) for some (small)* $\delta \neq 0$?

We are unable to answer this question analytically and present some numerical results using the Ambrosio–Tortorelli [5] approximation functional implemented via the free software *Freefem++*.

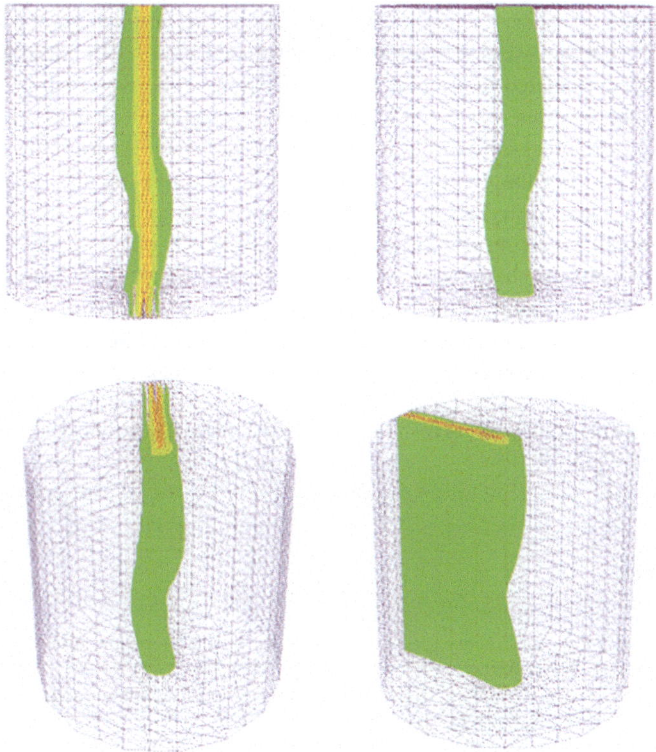

Fig. 4 The singular set of the solution with boundary datum u_δ on $\partial\mathscr{C}$ with $\delta = 1/2$. ©Elsevier, see [25]

In Fig. 4, we have presented some isovalues of the phase-field φ describing the discontinuity set P for the boundary data given by $u_{1/2}$. The profile of P looks like a twisted half-plane, which shows that $u_{1/2}$ may not be a minimizer. Moreover, the author thinks that the dependence of the function on the variable on the crack-front is related to the rotation of the co-normal tangent at the crack-front. We refer the interested reader to [25] for further details.

5 Models with Interaction Phenomena Between Phases

Considering the curvilinear geometry of cracks, the physical consistency would require to take into account interaction phenomena of the contacting crack faces. Following a suggestion by the referee, we would like to emphasize on the relation with those models.

The variational theory of nonlinear cracks subject to non-penetration and their growth under Griffith's brittle fracture was developed in [19, 20] and other works by the authors, extended to frictional models [3] and cohesive crack propagation [23, 26]. From the point of view of fracture criteria used in mechanics, the solution singularity at the crack-tip is of primary importance since determining the crack growth and path. For the nonlinear cracks, see the solution singularity analysis in [17, 18], the related J-integrals in [7, 21], and kink of cracks in [20]. In comparison, under Barenblatt's quasi-brittle fracture, the cohesive crack obeys no any singularity (see [23]). The present study relies on curvilinear cracks under stress-free conditions and the anti-plane simplification. The crack is not prescribed a priori and should be determined from a minimization of a Mumford–Shah function implying the total energy, based on the geometric concept of ε-close to a crack-tip singular solution.

Acknowledgements This work has been supported by the National Science Foundation of China (research grant no. 11650110437).

References

1. Ambrosio, L.: Existence theory for a new class of variational problems. Arch. Ration. Mech. Anal. **111**, 291–322 (1990)
2. Ambrosio, L., Fusco, N., Pallara, D.: Partial regularity of free discontinuity sets. II. Ann. Scuola Norm. Sup. Pisa Cl. Sci. (4) **24**, 39–62 (1997)
3. Ambrosio, L., Fusco, N., Pallara, D.: Functions of Bounded Variation and Free Discontinuity Problems. Oxford Mathematical Monographs. The Clarendon Press, Oxford University Press, New York (2000)
4. Ambrosio, L., Pallara, D.: Partial regularity of free discontinuity sets. I. Ann. Scuola Norm. Sup. Pisa Cl. Sci. (4) **24**, 1–38 (1997)
5. Ambrosio, L., Tortorelli, V.M.: On the approximation of free discontinuity problems. Boll. Un. Mat. Ital. B **7**(6), 105–123 (1992)
6. Andersson, J., Mikayelyan, H.: Regularity up to the crack-tip for the mumford-shah problem. In: First Report on Viscosity and Plasticity, 2nd edn., pp. 1–58 (2015). arXiv.org
7. Azegami, H., Ohtsuka, K., Kimura, M.: Shape derivative of cost function for singular point: evaluation by the generalized J integral. JSIAM Lett. **6**, 29–32 (2014)
8. Bonnet, A.: On the regularity of edges in image segmentation. Ann. Inst. H. Poincaré Anal. Non Linéaire **13**, 485–528 (1996)
9. Bonnet, A., David, G.: Cracktip is a global Mumford-Shah minimizer. Astérisque (274), Société Mathématique De France (2001)
10. David, G.: C^1-arcs for minimizers of the Mumford-Shah functional. SIAM J. Appl. Math. **56**, 783–888 (1996)
11. David, G.: Singular Sets of Minimizers for the Mumford-Shah Functional. Progress in Mathematics, vol. 333. Birkhäuser Verlag, Basel (2005)
12. De Giorgi, E., Carriero, M., Leaci, A.: Existence theorem for a minimum problem with free discontinuity set. Arch. Ration. Mech. Anal. **108**, 195–218 (1989)
13. De Lellis, C., Focardi, M.: Higher integrability of the gradient for minimizers of the 2d Mumford-Shah energy. J. Math. Pures Appl. **9**(100), 391–409 (2013)
14. De Philippis, G., Figalli, A.: Higher integrability for minimizers of the Mumford-Shah functional. Arch. Ration. Mech. Anal. **213**, 491–502 (2014)
15. Francfort, G.A., Marigo, J.-J.: Revisiting brittle fracture as an energy minimization problem. J. Mech. Phys. Solids **46**, 1319–1342 (1998)

16. Griffith, A.A.: The phenomena of rupture and flow in solids. Philos. Trans. R. Soc. Ser. A Math. Phys. Eng. Sci. **221**, 582–593 (1921)
17. Itou, H., Khludnev, A.M., Rudoy, E.M., Tani, A.: Asymptotic behaviour at a tip of a rigid line inclusion in linearized elasticity. ZAMM Z. Angew. Math. Mech. **92**, 716–730 (2012)
18. Itou, H., Kovtunenko, V.A., Tani, A.: The interface crack with Coulomb friction between two bonded dissimilar elastic media. Appl. Math. **56**, 69–97 (2011)
19. Khludnev, A.M., Kovtunenko, V.A.: Analysis of Cracks in Solids. WIT-Press, Southampton, Boston (2000)
20. Khludnev, A.M., Kovtunenko, V.A., Tani, A.: Evolution of a crack with kink and non-penetration. J. Math. Soc. Jpn. **60**, 1219–1253 (2008)
21. Khludnev, A.M., Ohtsuka, K., Sokoł owski, J.: On derivative of energy functional for elastic bodies with cracks and unilateral conditions. Q. Appl. Math. **60**, 99–109 (2002)
22. Koch, H., Leoni, G., Morini, M.: On optimal regularity of free boundary problems and a conjecture of De Giorgi. Commun. Pure Appl. Math. **58**, 1051–1076 (2005)
23. Kovtunenko, V.A., Sukhorukov, I.V.: Optimization formulation of the evolution problem of crack propagation under quasi-brittle fracture. Prikl. Mekh. Tekhn. Fiz. **47**, 107–118 (2006)
24. Lemenant, A.: A selective review on Mumford-Shah minimizers. Boll. Unione Mat. Ital. **9**, 69–113 (2016)
25. Lemenant, A., Mikayelyan, H.: Stationarity of the crack-front for the Mumford-Shah problem in 3D. J. Math. Anal. Appl. **462**, 1555–1569 (2018)
26. Leugering, G., Prechtel, M., Steinmann, P., Stingl, M.: A cohesive crack propagation model: mathematical theory and numerical solution. Commun. Pure Appl. Anal. **12**, 1705–1729 (2013)
27. Li, Z., Mikayelyan, H.: Fine numerical analysis of the crack-tip position for a Mumford-Shah minimizer. Interfaces Free Bound **18**, 75–90 (2016)
28. Mumford, D., Shah, J.: Optimal approximations by piecewise smooth functions and associated variational problems. Commun. Pure Appl. Math. **42**, 577–685 (1989)

Piecewise Constant Upwind Approximations to the Stationary Radiative Transport Equation

Hiroshi Fujiwara

Abstract We discuss piecewise constant approximations to the stationary radiative transport equation. Convergence of the proposed scheme is numerically studied with geometrically nonconformal and nonconvex polygonal meshes, and the results imply some extension of the conventional theoretical framework of the standard finite element method. An advantage of the proposed scheme in terms of reducing computational resources is also discussed in comparison with the finite difference method.

1 Introduction

In the present paper, we discuss numerical computation of the stationary radiative transport equation (RTE). RTE is a mathematical model to describe light propagation in terms of photon migration in random mediums such as biomedical tissue [11, 13], and its numerical treatment plays an essential role in the development of next-generation diagnostic methods using near-infrared light [10].

Let $\Omega \subset \mathbb{R}^2$ be a convex polygonal domain, $S = \{\xi \in \mathbb{R}^2 \,;\ |\xi| = 1\}$ be the unit circle, and $X = \Omega \times S$. Since the speed of light is sufficiently fast in comparison with the size of bodies, we focus on the following boundary value problem of RTE:

$$-\xi \cdot \nabla_x I(x,\xi) - (\mu_\mathrm{a} + \mu_\mathrm{s}) I(x,\xi) + \mu_\mathrm{s} \int_S p(x;\xi,\xi') I(x,\xi')\, d\sigma_{\xi'} = -q(x,\xi), \text{ in } X, \tag{1a}$$

$$I(x,\xi) = I_1(x,\xi), \text{ on } \Gamma_-, \tag{1b}$$

where $\nabla_x I = (\partial I/\partial x_1, \partial I/\partial x_2)$ and $d\sigma_\xi$ is the curve element on S. The function $I = I(x,\xi)$ represents a particle density at a position $x \in \Omega$ with a velocity $\xi \in S$.

H. Fujiwara (✉)
Graduate School of Informatics, Kyoto University, Yoshida-Honmachi,
Sakyo-ku, Kyoto 606-8501, Japan
e-mail: fujiwara@acs.i.kyoto-u.ac.jp

H. Itou et al. (eds.), *Mathematical Analysis of Continuum Mechanics and Industrial Applications III*, Mathematics for Industry 34,
https://doi.org/10.1007/978-981-15-6062-0_3

The boundary is decomposed into two parts:

$$\Gamma_{\pm} = \{(x, \xi) \in \partial\Omega \times S\,;\; \xi \cdot n(x) \gtrless 0\},$$

where $n(x)$ is the outward unit normal at x to $\partial\Omega$, and the inflow boundary condition (1b) is given on Γ_-. Functions $\mu_a = \mu_a(x)$ and $\mu_s = \mu_s(x)$ are called absorption and scattering coefficients, respectively, and the scattering kernel $p(x; \xi, \xi')$ is the conditional probability of changing velocity from ξ' to ξ scattered at x. The inhomogeneous term $q = q(x, \xi)$ is an internal particle source.

Even though Ω is a two-dimensional domain, the unknown function I depends on $(x, \xi) \in \Omega \times S$. It means that the straightforward discretization of RTE leads a three-dimensional large-scale problem. To save computational resources, the diffusion approximation has been frequently used by introducing a new function $\Phi(x) = \int_S I(x, \xi) d\sigma_\xi$ [2, 14], which is the first mode of the Fourier series (or the spherical harmonic series in three dimensions) of $I(x, \xi)$. Since the diffusion approximation computes only $\Phi(x)$ without the dependency on ξ, additional algorithms are required in order to recover directivity of light propagation. On the other hand, the Monte Carlo method is one of standard numerical procedures in biomedical studies [3, 15]. It involves relatively large statistical errors, and thus accurate and reliable numerical methods have been still under development.

Recently, the discrete ordinate discontinuous Galerkin method has been developed as an effective method [8, 9] based on regularity of the exact solution to RTE [1]. It is formulated by the discontinuous Galerkin method and the collocation method w.r.t. x and ξ, respectively. Its convergence was proved in [9], where they mainly focused on spatial discretization errors with sufficiently accurate discretization for ξ.

In this paper, we adopt the piecewise constant approximation w.r.t. both x and ξ and construct another discretization scheme similar to finite volume methods, since RTE is based on a balance law of particles. We also study its convergence by numerical experiments.

Throughout the paper, the following conditions are assumed.

1. Ω is a convex polygonal domain in $\mathbb{R}^2$.
2. $q \in L^2(X)$, $I_1 \in L^2(\Gamma_-)$.
3. $p \in L^\infty(\Omega \times S \times S)$, $p \geq 0$, $p(x; \xi, \xi') = p(x; \xi', \xi)$, and $\int_S p(x; \xi, \xi')\, d\sigma_{\xi'} = 1$, a.e. x, ξ.
4. $\mu_t = \mu_s + \mu_a$, μ_s, $\mu_a \in L^\infty(\Omega)$, $\mu_s \geq 0$, and there exists a positive number $\mu_a^- \in \mathbb{R}$ such that $\mu_a \geq \mu_a^-$.
5. There exists a unique solution to RTE in $V = \{f\,;\; f, \nabla_x f, \partial_\xi f \in L^2(X)\}$, where $\partial_\xi = \partial/\partial\theta$ is calculated by the identification $\xi = (\cos\theta, \sin\theta) \in S$ with $\theta \in [0, 2\pi)$.

We give a remark on terminology of "convergence" in the present paper. We shall discuss error behaviors with a mesh-dependent norm. Strictly speaking, "convergence" with the mesh-dependent norm is abuse since error is measured with different norms. However, we use it for clarity.

2 Piecewise Constant Upwind Approximations to RTE

In this section, we derive a discretization scheme to RTE by employing the piecewise constant approximation w.r.t. both x and ξ variables.

First, we introduce some notations to describe the proposed scheme. Let $\mathscr{K}$ be a polygonal mesh of Ω, i.e., each $\kappa \in \mathscr{K}$ is a nonempty polygonal domain and

$$\kappa, \tau \in \mathscr{K}, \kappa \neq \tau \Rightarrow \kappa \cap \tau = \emptyset, \quad \text{and} \quad \overline{\Omega} = \bigcup_{\kappa \in \mathscr{K}} \overline{\kappa}.$$

Subsets of $\partial\kappa$ for a fixed $\xi \in S$ are denoted by

$$\partial\kappa_{\xi}^{\pm} = \{x \in \partial\kappa \ ; \ \xi \cdot n_{\kappa}(x) \gtrless 0\},$$

where $n_{\kappa}(x)$ is the outer unit normal to $\partial\kappa$. Define the set of adjacent elements of $\kappa \in \mathscr{K}$ as

$$\partial\Lambda_{\kappa} = \{\tau \in \mathscr{K} \ ; \ \tau \neq \kappa, |\partial\tau \cap \partial\kappa| \neq 0\}.$$

The unit circle $S = [0, 2\pi)$ is split into $\omega_1, \ldots, \omega_N$ where each ω_n is identified with the open interval $(\theta_{n-1}, \theta_n) \mod 2\pi$.

Let $V_h = V_h(\mathscr{K} \times \{\omega_n\})$ be the set of $\mathscr{K} \times \{\omega_n\}$-piecewise (element-wise) constant functions, that is, the restriction of a function in V_h on each $\kappa \times \omega \in \mathscr{K} \times \{\omega_n\}$ is constant. We introduce a vector space $V + V_h = \{v + v_h \ ; \ v \in V, v_h \in V_h\}$.

For $u \in V + V_h$ and $x \in \partial\kappa$, we note that the directional limits

$$u^{\pm}(x, \xi) = \lim_{\varepsilon \to +0} \tilde{u}(x \pm \varepsilon\xi, \xi)$$

exist in the trace sense if $n_{\kappa}(x) \cdot \xi \neq 0$, where $\tilde{u}$ is the zero extension of u:

$$\tilde{u}(x, \xi) = \begin{cases} u(x, \xi), & x \in \Omega, \\ 0, & \text{otherwise.} \end{cases}$$

Now we discretize RTE using above notations. Suppose that $I \in V$ satisfies (1a). Integration of (1a) on $\kappa \times \omega \in \mathscr{K} \times \{\omega_n\}$, and Green's formula yield

$$\int_{\omega}\int_{\partial\kappa\cap\Omega} (\xi\cdot n(x))I(x,\xi)\,\mathrm{d}\ell\,\mathrm{d}\sigma_\xi + \int_{\Gamma_+\cap(\partial\kappa\times\omega)} (\xi\cdot n(x))I(x,\xi)\,\mathrm{d}\ell\,\mathrm{d}\sigma_\xi$$
$$+\int_{\kappa\times\omega}\left\{\mu_{\mathrm{t}}(x)I(x,\xi)-\mu_{\mathrm{s}}(x)\int_S p(x;\xi,\xi')I(x,\xi')\,\mathrm{d}\sigma_{\xi'}\right\}\mathrm{d}x\,\mathrm{d}\sigma_\xi$$
$$=\int_{\kappa\times\omega} q(x,\xi)\,\mathrm{d}x\,\mathrm{d}\sigma_\xi + \int_{\Gamma_-\cap(\partial\kappa\times\omega)} \big|\xi\cdot n(x)\big|I_1(x,\xi)\,\mathrm{d}\ell\,\mathrm{d}\sigma_\xi, \qquad (2)$$

where $\mathrm{d}\ell$ is the line element along $\partial\kappa$. We adopt the upwind approximation $I\approx I^-$ on $\partial\kappa$ in the first and the second terms. Further, we apply the piecewise constant approximation $I\approx I_h\in V_h$. These two approximations lead

$$\int_{\omega}\int_{\partial\kappa\cap\Omega} (\xi\cdot n)I_h^-(x,\xi)\,\mathrm{d}\ell\,\mathrm{d}\sigma_\xi + \int_{\Gamma_+\cap(\partial\kappa\times\omega)} (\xi\cdot n)I_h^-(x,\xi)\,\mathrm{d}\ell\,\mathrm{d}\sigma_\xi$$
$$+\int_{\kappa\times\omega}\left\{\mu_{\mathrm{t}}I_h-\mu_{\mathrm{s}}\int_S pI_h(x,\xi')\,\mathrm{d}\sigma_{\xi'}\right\}\mathrm{d}x\,\mathrm{d}\sigma_\xi$$
$$=\int_{\kappa\times\omega} q\,\mathrm{d}x\,\mathrm{d}\sigma_\xi + \int_{\Gamma_-\cap(\partial\kappa\times\omega)} |\xi\cdot n|I_1\,\mathrm{d}\ell\,\mathrm{d}\sigma_\xi.$$

Since I_h is constant on each $\kappa\times\omega$, $I_h|_{\kappa\times\omega}=I_{\kappa,\omega}$, the following numerical scheme is obtained: for any $\kappa\times\omega\in\mathscr{K}\times\{\omega_n\}$,

$$\sum_{\tau\in\partial\Lambda_\kappa}\left(\int_{\omega}\int_{\partial\tau\cap\partial\kappa_\xi^-\cap\Omega} |\xi\cdot n(x)|\,\mathrm{d}\ell\,\mathrm{d}\sigma_\xi\right)(I_{\kappa,\omega}-I_{\tau,\omega})$$
$$+\left(\int_{\omega}\int_{\partial\kappa_\xi^-\cap\partial\Omega} |\xi\cdot n(x)|\,\mathrm{d}\ell\,\mathrm{d}\sigma_\xi\right)I_{\kappa,\omega}$$
$$+|\omega|\left(\int_{\kappa}\mu_{\mathrm{t}}(x)\,\mathrm{d}x\right)I_{\kappa,\omega}-\sum_{\omega'}\left(\int_{\kappa\times\omega_\xi\times\omega'_{\xi'}}\mu_{\mathrm{s}}(x)p(x;\xi,\xi')\,\mathrm{d}x\,\mathrm{d}\sigma_\xi\,\mathrm{d}\sigma_{\xi'}\right)I_{\kappa,\omega'}$$
$$=\int_{\kappa\times\omega} q(x,\xi)\,\mathrm{d}x\,\mathrm{d}\sigma_\xi + \int_{\Gamma_-\cap(\partial\kappa\times\omega)} |\xi\cdot n(x)|I_1(x,\xi)\,\mathrm{d}\ell\,\mathrm{d}\sigma_\xi. \qquad (3)$$

Since the scheme (3) is considered as a finite volume method, the next lemma follows directly from finite difference analysis in [7] as a generalized finite difference method.

Lemma 1 *We assume that* $\sup_{\substack{x\in\Omega\\ \omega\in\{\omega_n\}}}\left(|\omega|\sup_{\xi,\xi'\in\omega} p(x;\xi,\xi')\right)\le 1$. *Then, there uniquely exists an elementary matrix of interexchanging rows which transforms the coefficient matrix into a strictly diagonally dominant one.*

It follows from the above lemma immediately that there exists a unique solution $I_h \in V_h$ to (3). It is also a direct consequence that Gauss–Seidel iteration gives I_h to the system of linear equations (3).

3 Numerical Study for Convergence of the Proposed Scheme

In this section, we exhibit some numerical experiments to verify convergence of the proposed scheme.

Let Ω be the unit square and $\mu_s = 1$, $\mu_a = 0.001$. Similarly as an example in [9], we use the Poisson kernel

$$p(x;\xi,\xi') = \frac{1}{2\pi}\frac{1-g^2}{1-2g\cos(\theta-\theta')+g^2}, \quad g = 0.9, \tag{4}$$

as the scattering kernel. The inhomogeneous term and the boundary condition are

$$\begin{aligned} q(x_1,x_2,\theta) &= 2\pi x_1\xi_1^2\cos(\pi x_1^2)\cos(2\pi x_2^2)\\ &\quad - 4\pi x_2\xi_1\xi_2\sin(\pi x_1^2)\sin(2\pi x_2^2)\\ &\quad + (\mu_a + (1-g)\mu_s)\xi_1\sin(\pi x_1^2)\cos(2\pi x_2^2), \end{aligned}$$

and

$$I_1(x_1,x_2,\theta) = \begin{cases} 0, & x_1 = 0 \text{ or } x_1 = 1,\\ \xi_1\sin(\pi x_1^2), & x_2 = 0 \text{ or } x_2 = 1. \end{cases}$$

This setting admits the exact solution given by

$$I(x_1,x_2,\theta) = \xi_1\sin(\pi x_1^2)\cos(2\pi x_2^2).$$

Since the proposed scheme is a kind of finite volume methods, the discontinuous Galerkin techniques are applicable to prove convergence [6, 9]. Therefore, numerical error is measured in the sense of the standard $L^2(X)$ norm and a mesh-dependent norm with $\mathscr{K}_h \times \{\omega_n\}$ as

$$\|u\|_{h,N}^2 = \|u\|_{L^2(X)}^2 + \|u^+ - u^-\|_{L_f^2(\mathscr{E}_h^i\times S)}^2 + \|u^-\|_{L_f^2(\Gamma_+)}^2 + \|u^+\|_{L_f^2(\Gamma_-)}^2,$$

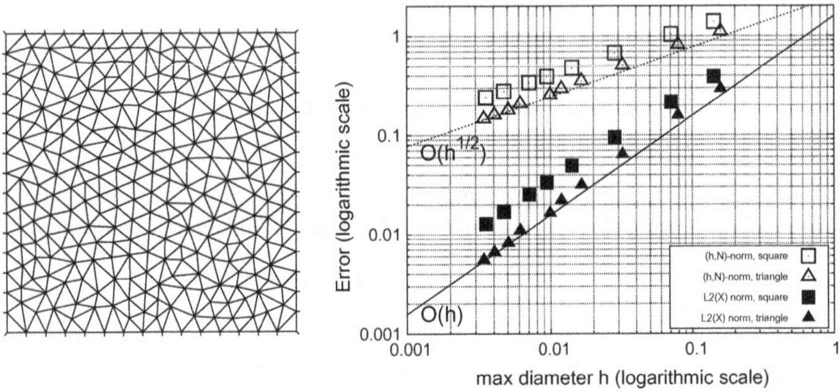

Fig. 1 Example of triangle mesh ($h \approx 0.109$) and error behaviors with triangle meshes and square meshes

where

$$\|u\|^2_{L^2_f(e\times\omega)} = \int_{e\times\omega} |\xi \cdot n(x)|\, u(x,\xi)^2 \,\mathrm{d}\ell\, \mathrm{d}\sigma_\xi, \quad e \subset \mathscr{E}_h \text{ and } \omega \subset S.$$

Sets of spatial edges in $\mathscr{K}_h$ are denoted by

$$\mathscr{E}_h = \bigcup_{\kappa\in\mathscr{K}_h} \partial\kappa, \quad \mathscr{E}_h^i = \mathscr{E}_h \setminus \partial\Omega = \mathscr{E}_h \cap \Omega.$$

It should be noted that $\|\cdot\|_{h,N}$ is used so as to detect jumps across edges [6], and that it can be applied to $u \in V + V_h$.

Three kinds of spatial meshes are examined to discuss convergence of the proposed scheme (3). The symbol h indicates the maximum of diameters of the spatial mesh as $h = \max\{\mathrm{diam}(\kappa)\,;\, \kappa \in \mathscr{K}_h\}$.

Example 1 (*Regular Mesh*). *Regular* triangle and square meshes are used. Equi-spaced $4N$ vertices are placed on $\partial\Omega$ and $\omega_n = \big((n-1)\Delta\theta, n\Delta\theta\big)$ with $\Delta\theta = 2\pi/N$. An example of triangle meshes is depicted in Fig. 1 (left).

The numerical results suggest that

$$\|I - I_h\|_{h,N} = O(h^{1/2}) \quad \text{and} \quad \|I - I_h\|_{L^2(X)} = O(h).$$

Example 2 (*Geometrically Nonconformal Mesh*). Meshes are generated as follows: Ω is decomposed into $N \times N$ squares, and some squares are randomly split into either rectangles, squares, or triangles (Fig. 2 left). The interval ω_n is the same as Example 1. This yields a geometrically nonconformal mesh [6], where some vertices lie on edges of other elements. Our numerical experiments show convergence rates

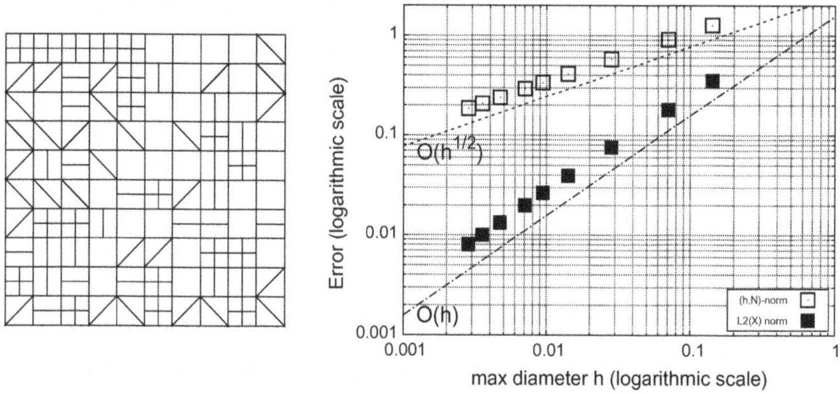

Fig. 2 Example of geometrically nonconformal meshes ($h = \sqrt{2}/10 \approx 0.141$) and error behaviors

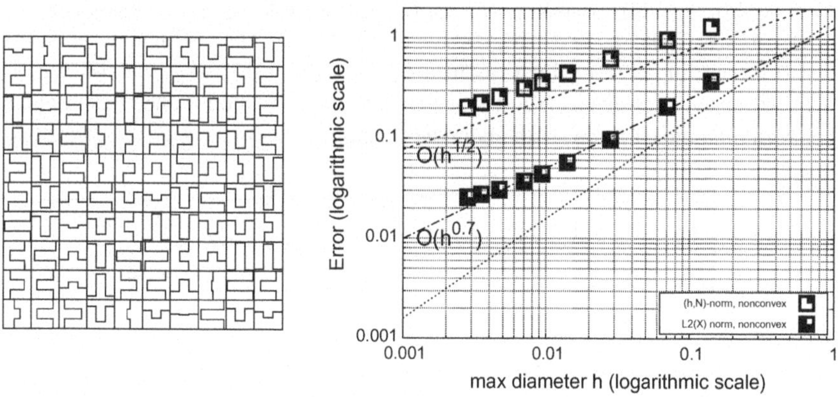

Fig. 3 Example of nonconvex polygonal meshes ($h \approx \sqrt{2}/10 \approx 0.141$) and error behaviors

similar to Example 1. We remark that nonconformal meshes enable us to generate graded meshes and mesh refinements without any difficulties, although they are treated as a *forbidden* situation [5] in theory of the finite element analysis.

Example 3 (*Nonconvex Mesh*). Finally, we adopt nonconvex meshes, which is also discarded in the conventional theoretical framework of the finite element method since the interpolation error analysis on each element is based on the homogeneity argument [4]. Figure 3 shows an example of meshes (left) and numerical errors (right). The convergence rate of numerical errors in $\|\cdot\|_{h,N}$ is $O(h^{1/2})$, while that in $L^2(X)$-norm is not $O(h)$ but $O(h^{0.7})$.

These numerical examples suggest that

$$\|I - I_h\|_{h,N} = O(h^{1/2})$$

even for geometrically nonconformal and nonconvex meshes, while mesh convexity is required to achieve convergence as

$$\|I - I_h\|_{L^2(X)} = O(h).$$

4 Efficiency of Graded Meshes

One of the difficulties in numerical computation of RTE is its requirement of huge computational resources. In this section, we discuss the efficiency of the proposed scheme in comparison with the finite difference method (FDM) [7, 12] from the quantitative viewpoint.

Let $\Omega = (-10, 10) \times (0, 20)$. We consider two polygonal inclusions depicted with gray in Fig. 4; vertices of the pentagon are $(-3.28, 12.42)$, $(2.72, 14.42)$, $(0.72, 16.20)$, $(-3.50, 16.44)$, $(-5.54, 14.32)$ and those of the quadrilateral are $(-5.18, 4.34)$, $(6.08, 4.78)$, $(7.08, 10.84)$, $(2.42, 12.82)$. The coefficients are set as $\mu_s = 0.109$ and $\mu_a = 0.008$ in the inclusions, and $\mu_s = 1.09$ and $\mu_a = 0.08$ outside the inclusions. We pose $q \equiv 0$ and p as (4). Assuming light emittance from optical fiber with diameter 0.1, we set the boundary condition as

$$I_1(x, \theta) = \begin{cases} \left\{1 - \left(\dfrac{x}{r}\right)^2\right\}^3 \dfrac{1}{\sqrt{2\pi}\sigma} \exp\left(-\dfrac{(\theta - \pi/2)^2}{2\sigma^2}\right), & |x_1| < r \text{ and } x_2 = 0, \\ 0, & \text{otherwise} \end{cases}$$

with $r = 0.05$, $\sigma = 0.2$.

Figure 5 shows the profile of the total light intensity $\Phi(x) = \int_S I(x, \xi)\, d\sigma_\xi$ on Ω, and Fig. 6 shows its decay along $x_1 = 0$ with $\Delta x_1 = \Delta x_2$ and $\Delta\theta = 2\pi/360$. The results suggest that the numerical solution by FDM converges as Δx_i tends to zero.

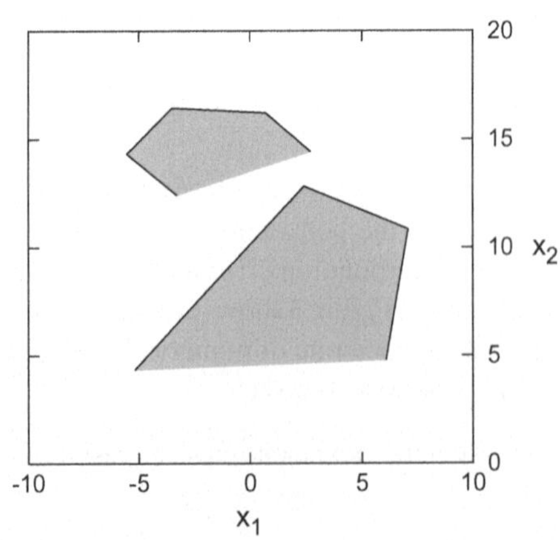

Fig. 4 Polygonal inclusions filled with gray

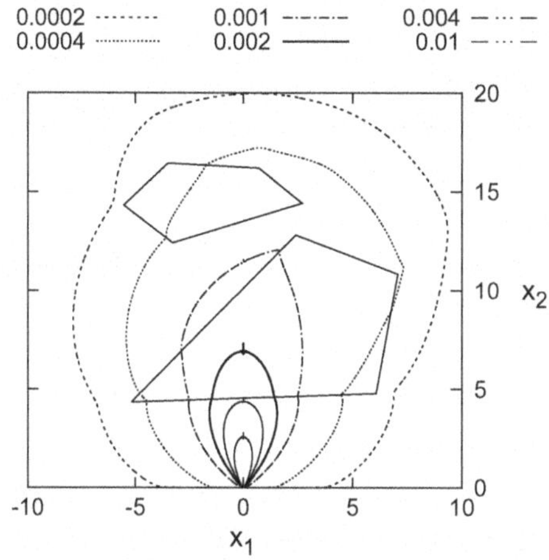

Fig. 5 Contour lines of numerical solution of total light intensity $\Phi_h(x_1, x_2)$ with $\Delta x_i = 0.01$, $\Delta\theta = 2\pi/360$

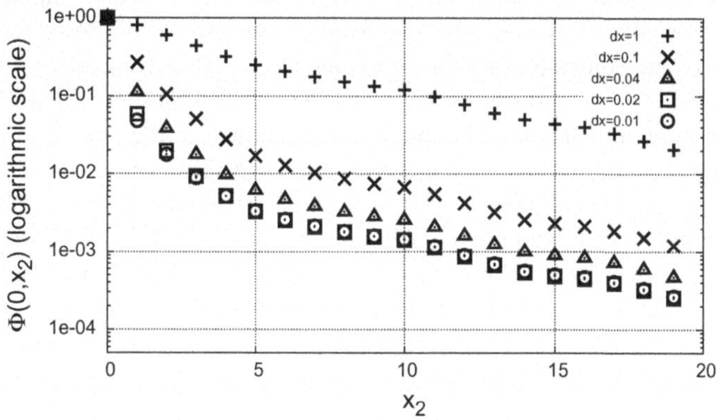

Fig. 6 Decay of $\Phi(0, x_2)$ along the line $x_1 = 0$, which shows that the numerical results with $\Delta x_i = 0.02$ (□) and those with $\Delta x_i = 0.01$ (○) show good agreements

In particular, we can conclude that FDM solutions with $\Delta x_i = 0.02$ and $\Delta x_i = 0.01$ are reliable from the quantitative viewpoint in this example. On the other hand, the number of unknowns for $\Delta x_i = 0.01$ and $\Delta\theta = 2\pi/360$ is approximately 1.4 billions, which corresponds to 10GB in the double precision. The computational time is about 5.5 hours with 68 threads parallel computation (9200 iterations with Gauss–Seidel method to reduce residual less than 10^{-12}) on XeonPhi 7250 processor. This means that for reliable computation with FDM requires huge computational resources.

In order to reduce computational resources, we pay attention to concentration of the boundary condition I_1 at the origin, which corresponds to light emittance from

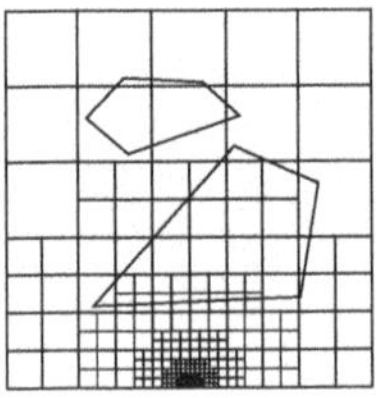

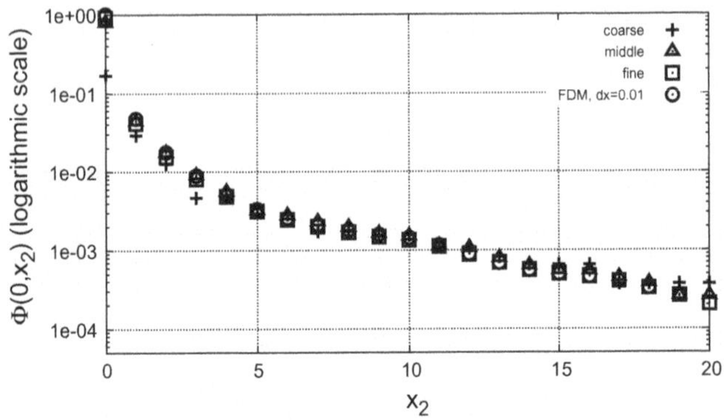

Fig. 7 Example of nonconformal graded polygonal mesh ($\#\mathscr{K} = 320$) and numerical results $\Phi(0, x_2)$ with the proposed method (+, △, □) and FDM ($\Delta x_i = 0.01$, ○)

a thin optical fiber in practical situations. It is natural to introduce a graded mesh depicted in Fig. 7 (left). We note that the mesh contains geometrically nonconformal and nonconvex polygonal elements, which are accepted by the proposed scheme (3).

Figure 7 (right) illustrates decay of $\Phi(x)$ along $x_1 = 0$ with coarse ($\#\mathscr{K} = 320$), middle ($\#\mathscr{K} = 5463$), and fine ($\#\mathscr{K} = 211933$) graded meshes. Figure 7 demonstrates their good agreements. The number of unknowns in the fine mesh case with $\Delta\theta = 2\pi/360$ is 76 millions, which corresponds to 582MB (5.3% of FDM) in the double precision. The computational time is about 10 minutes (3.1% of FDM, 1560 iterations with Gauss–Seidel method to reduce residual less than 10^{-12}) on the same computational environment. From this experiments, it is clearly concluded that the proposed method is quite efficient to reduce computational resources.

5 Concluding Remarks

The proposed scheme derived from the piecewise constant upwind approximation shows two types of error convergence: $O(h)$ with the standard L^2-norm and $O(h^{1/2})$ with the mesh-dependent norm even for geometrically nonconformal polygonal mesh. The difference is that the former requires convexity of meshes, while the latter holds on nonconvex meshes. A proof of convergence will be given in a subsequent paper.

Acknowledgments The author thanks Professor Yuusuke Iso, Professor Nobuyuki Higashimori, and Professor Naoya Oishi for their valuable comments. This work was partially supported by JSPS KAKENHI Grant Numbers 16H02155, 18K18719, and 18K07712, 18K03436.

References

1. Agoshkov, V.: Boundary Value Problems for Transport Equations. Birkhäuser, Boston (1998)
2. Arridge, S.R.: Optical tomography in medical imaging. Inverse Prob. **15**, R41–R93 (1999)
3. Boas, D.A., Culver, J.P., Stott, J.J., Dunn, A.K.: Three dimensional Monte Carlo code for photon migration through complex heterogeneous media including the adult human head. Opt. Exp. **10**, 159–170 (2002)
4. Brenner, S.C., Scott, L.R.: The Mathematical Theory of Finite Element Methods, 3rd edn. Springer, Berlin (2008)
5. Ciarlet, P.G.: The Finite Element Method for Elliptic Problems, 2nd edn. SIAM (2002)
6. Ern, A., Guermond, J.L.: Theory and Practice of Finite Elements. Springer, New York (2004)
7. Fujiwara, H.: Numerical analysis of upwind finite difference scheme to the stationary radiative transport equation for light propagation in biomedical tissue (in Japanese). JASCOME Rev. **2017**, 21–32 (2017)
8. Guermond, J.L., Kanschat, G., Ragusa, J.C.: Discontinuous Galerkin for the radiative transport equation. In: Feng, X., et al. (eds.) Recent Developments in Discontinuous Galerkin Finite Element Methods for Partial Differential Equations. Springer, Switzerland (2014)
9. Han, W., Huang, J., Eichholz, J.A.: Discrete-ordinate discontinuous Galerkin methods for solving the radiative transfer equation. SIAM J. Sci. Comput. **32**, 477–497 (2010)
10. Jöbsis, F.F.: Noninvasive, infrared monitoring of cerebral and myocardial oxygen sufficiency and circulatory parameters. Science **198**, 1264–1267 (1977)
11. Ishimaru, A.: Theory and application of wave propagation and scattering in random media. Proc. IEEE **65**, 1030–1061 (1977)
12. Klose, A.D., Netz, U., Beuthan, J., Hielscher, A.H.: Optical tomography using the time-independent equation of radiative transfer—part 1: forward model. J. Quant. Spectrosc Radiat. Transfer **72**, 691–713 (2002)
13. Nagirner, D.I.: V.V. Sobolev and analytical radiative transfer theory. In: Grinin, V. et al. (eds.) Radiation Mechanisms of Astrophysical Objects, pp. 3–28. Edit Print Publishing House (2017)
14. Yamada, Y., Okawa, S.: Diffuse optical tomography: present status and its future. Opt. Rev. **21**, 185–205 (2014)
15. Wang, L., Jacques, S.L., Zheng, L.: MCML–Monte Carlo modeling of light transport in multi-layered tissues. Comput. Methods Programs Biomed. **47**, 131–146 (1995)

Elasticity

The Mechanics and Mathematics of Bodies Described by Implicit Constitutive Equations

K. R. Rajagopal

Abstract In this short paper, I provide a brief discussion of the rationale and need for implicit constitutive relations for describing the response of many real materials. Classical theories to describe the response of materials such as the Cauchy theory of elasticity and the Navier–Stokes equations turn the demands of causality on its head and provide an expression for the Cauchy stress in terms of kinematical variables. It would be more reasonable instead to provide an expression for an appropriate kinematical variable in terms of the stress, if possible. However, in some instances such a specification might not be possible and we might have to resort to the prescription of an implicit relationship between the stress, its various frame-indifferent time derivatives, the appropriate kinematical variables and their frame-indifferent time derivative, as well as other relevant variables that influence the response of the body.

1 Introduction

In classical theories that have been developed to describe the response of bodies one usually finds an expression for the stress in terms of kinematical quantities. The approximations of linearized elasticity and linearized viscoelasticity are different in that one finds both expressions for the kinematical quantity in terms of the stress as well as the stress in terms of kinematical quantities. This is not a consequence of the employers of such models having understood the implications of causality or the philosophical implications of the choice of the different forms of constitutive expressions. Most times it is based merely with the ease in using these forms in different situations, and the invertibility of constitutive relations providing one with the mathematical option of using whichever approach is preferable.

Though referred to at times as constitutive relations, in classical theories of elasticity and fluid mechanics what one finds are constitutive functions that provide

K. R. Rajagopal (✉)
Texas A&M University, College Station, Texas 77843, USA
e-mail: krajagopal@tamu.edu

H. Itou et al. (eds.), *Mathematical Analysis of Continuum Mechanics and Industrial Applications III*, Mathematics for Industry 34,
https://doi.org/10.1007/978-981-15-6062-0_4

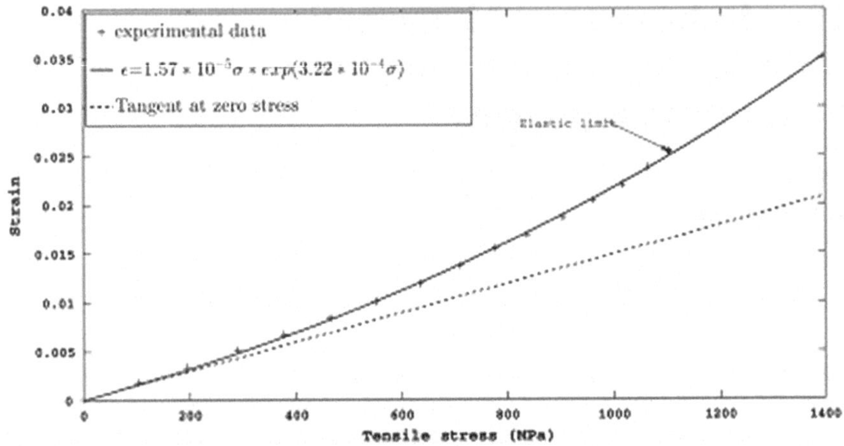

Fig. 1 Response of gum metal at small strains

Fig. 2 Shear stress versus shear rate in the experiments by Boltenhagen et al. [1] for Tris (2-hydroxyethyl) tallowalkyl ammonium acetate (TTAA) surfactant dissolved in water containing sodium salicylate (NaSal)

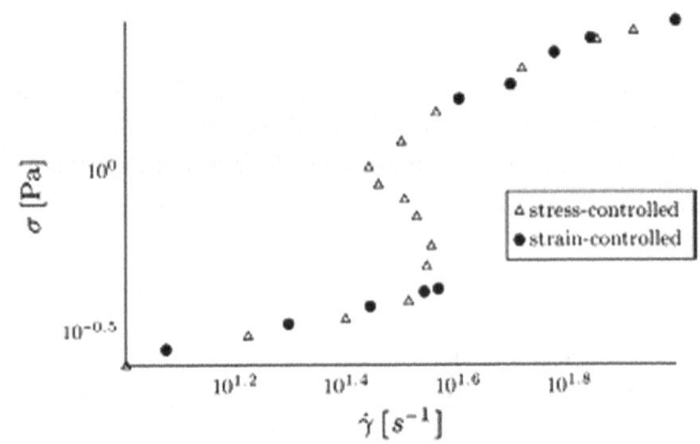

expressions for the stress in terms of kinematical variables, the density and deformation gradient in elasticity and the density and velocity gradient in fluid mechanics. One however finds implicit relationships for modeling the inelastic response of solids and the viscoelastic behavior of fluids.

With regard to elastic bodies, many metallic alloys respond in a nonlinear fashion even within the context of small strains (see Fig. 1) and such nonlinear behavior cannot be described within the context of Cauchy elasticity (Green elasticity is a subset of Cauchy elasticity), such response can however be described with great felicity within the context of linearizations of implicit models to describe elastic response. Similarly, with respect to the response of fluids, many colloids and suspensions exhibit the response portrayed in Fig. 2. Such response cannot be cast within the framework of the theory of simple fluids, one needs an implicit constitutive theory to describe the same. The development of implicit constitutive relations is motivated by the above experimental results as well as many other such examples too numerous to list in detail.

Before getting into a discussion of implicit constitutive models for describing the response of bodies, it is worthwhile to discuss the relevance of causality in classical Newtonian mechanics. The following comments of Newton [29] unequivocally identifies the force as the cause and the motion as the effect:

> The causes by which true and relative motion are distinguished, one from the other, are the forces impressed upon bodies to generate motion.

And

> The alteration of motion is ever proportional to the motive force impressed and is made in the direction of the right line in which that force is impressed.

Similar sentiments are expressed by numerous other pioneers of mechanics and we shall not quote them. More recently, Truesdell observes that

> A constitutive equation is a relation between forces and motions. In popular terms, force is applied to a body cause it to undergo a motion, and the motion caused differs according to the nature of the body. In continuum mechanics the forces of interest are contact forces, which are specified by the stress tensor **T**.

While it might be ideal, if possible, to express the effect, in our case the motion, in terms of the cause, the force (and consequently the stress), and thus an expression for the kinematical quantity in terms of the stress, we might not be able, in a complicated system with numerous causes and effects to even delineate those that are to be deemed as causes and those that are to be presumed as effects. The English philosopher David Hume recognized that while the connection between cause and effect is necessary to take into consideration in the explanation of phenomena, he observed that one cannot necessarily associate a specific cause with regard to a specific effect. The following remarks make his position clear on this issue:

> When we look about us toward external objects, and consider the operation of causes, we are never able, in a single instance, to discover any power or necessary connexion- - -.We only find, that the one does actually, in fact, follow the other." —"But when one particular species of event has always, in all instances, been conjoined with another, we make no longer any scruple of foretelling one upon the appearance of the other...We then call that one object, Cause; the other, Effect. We suppose that there is some connexion between them; some power in the one, by which it infallibly produces the other, and operates with the greatest certainty and strongest necessity.

Hume felt that the best that one could do is an implicit relationship between all the quantities that seem relevant to a phenomenon, those that are possibly causes and those that are possibly effects. We shall not expound on the philosophical issues of causality. The notion of force being the cause is an important part of Kantian philosophy.

2 Kinematics

Let $\mathbf{x}$ denote the current position of a particle which is at $\mathbf{X}$ in a stress-free reference configuration. Let $\mathbf{x} = \boldsymbol{\chi}(\mathbf{X}, t)$ denote the motion of a particle and let us denote the displacement by

$$\mathbf{u} := \mathbf{x} - \mathbf{X}. \tag{1}$$

The displacement gradients $\frac{\partial \mathbf{u}}{\partial \mathbf{X}}$ and $\frac{\partial \mathbf{u}}{\partial \mathbf{x}}$ are given by

$$\frac{\partial \mathbf{u}}{\partial \mathbf{X}} = \nabla_{\mathbf{X}} \mathbf{u} = \mathbf{F} - \mathbf{1}, \tag{2}$$

and

$$\frac{\partial \mathbf{u}}{\partial \mathbf{x}} = \nabla_{\mathbf{x}} \mathbf{u} = \mathbf{1} - \mathbf{F}^{-1}, \tag{3}$$

where $\mathbf{F}$ is the deformation gradient defined through

$$\mathbf{F} = \frac{\partial \chi}{\partial \mathbf{X}}. \tag{4}$$

The velocity is defined through

$$\mathbf{v} = \frac{\partial \chi}{\partial t}, \tag{5}$$

the velocity gradient $\mathbf{L}$, and its symmetric part $\mathbf{D}$ and its skew part $\mathbf{W}$ through

$$\mathbf{L} = \frac{\partial \mathbf{v}}{\partial \mathbf{x}}, \quad \mathbf{D} = \frac{1}{2}\left[\mathbf{L} + \mathbf{L}^T\right], \quad \mathbf{W} = \frac{1}{2}\left[\mathbf{L} - \mathbf{L}^T\right]. \tag{6}$$

The Cauchy-Green tensors $\mathbf{B}$ and $\mathbf{C}$ are defined through

$$\mathbf{B} := \mathbf{F}\mathbf{F}^T, \quad \mathbf{C} := \mathbf{F}^T\mathbf{F}. \tag{7}$$

The Green-St.Venant strain $\mathbf{E}$ and the Almansi-Hamel strain $\mathbf{e}$ are defined through

$$\mathbf{E} := \frac{1}{2}(\mathbf{C} - \mathbf{1}) = \frac{1}{2}\left[\left(\frac{\partial \mathbf{u}}{\partial \mathbf{X}}\right) + \left(\frac{\partial \mathbf{u}}{\partial \mathbf{X}}\right)^T + \left(\frac{\partial \mathbf{u}}{\partial \mathbf{X}}\right)^T\left(\frac{\partial \mathbf{u}}{\partial \mathbf{X}}\right)\right], \tag{8}$$

and

$$\mathbf{e} := \frac{1}{2}\left(\mathbf{1} - \mathbf{B}^{-1}\right) = \frac{1}{2}\left[\left(\frac{\partial \mathbf{u}}{\partial \mathbf{x}}\right) + \left(\frac{\partial \mathbf{u}}{\partial \mathbf{x}}\right)^T - \left(\frac{\partial \mathbf{u}}{\partial \mathbf{x}}\right)^T\left(\frac{\partial \mathbf{u}}{\partial \mathbf{x}}\right)\right]. \tag{9}$$

When the gradient of displacement gradient is sufficiently small in the sense of the Frobenius norm, namely, that for all points belonging to the body for all times,

$$\max \left\| \frac{\partial \mathbf{u}}{\partial \mathbf{X}} \right\| = O(\delta), \quad \delta \ll 1, \tag{10}$$

then

$$\mathbf{E} \approx \mathbf{e} \approx \boldsymbol{\varepsilon} + O(\delta^2), \tag{11}$$

where

$$\boldsymbol{\varepsilon} = \frac{1}{2}\left[\left(\frac{\partial \mathbf{u}}{\partial \mathbf{x}}\right) + \left(\frac{\partial \mathbf{u}}{\partial \mathbf{x}}\right)^T\right], \tag{12}$$

is referred to as the linearized strain.
The above kinematical definitions are sufficient for our purpose. The interested reader can find more details concerning kinematics in Truesdell [56].

3 Implicit Theories for Elastic Bodies

Before getting into a discussion for implicit constitutive relations for elastic bodies, we will consider briefly classical constitutive theories to describe the response of elastic bodies. We start with a review of the classical linearized approximation for the response of elastic bodies.

3.1 The Linearized Approximation of Elasticity for Isotropic Bodies

The classical constitutive approximation for the response of an isotropic linearized elastic body takes the form:

$$\mathbf{T} = \lambda(tr\boldsymbol{\varepsilon})\mathbf{1} + 2\mu\boldsymbol{\varepsilon}, \tag{13}$$

where λ and μ are the Lamé constants. The constitutive expression (13) can be expressed as

$$\boldsymbol{\varepsilon} = \frac{1}{E}\left[(1+\nu)\mathbf{T} - \nu(tr\mathbf{T})\mathbf{1}\right], \tag{14}$$

where E is Young's modulus and ν is Poisson's ratio. Young [60] introduced a modulus of elasticity but this modulus is different from that which bears his name.

It is also worth recalling that Young [59] believed that the modulus of elasticity is different for compression and tension.

The expression (14) expresses the effect, the kinematical quantity (in this case the strain) in terms of the stress. The two material moduli that appear in the constitutive relation both have clear physical meaning and can be measured easily through the conduct of simple experiments. On the other hand, in the constitutive expression for the stress while the material modulus μ has clear physical meaning and can be measured easily the same cannot be said for the material modulus λ.

Both the models discussed above are explicit models, the former giving an expression for the stress in terms of the linearized strain while the latter provides an explicit expression for the linearized strain in terms of the stress. The latter model meets the demands of causality and also leads to meaningful measurable material properties by means of which the body can be characterized. On the other hand, the model (13) turns the demands of causality topsy-turvy and uses a material modulus that defies clear interpretation in the representation. It is yet the preferred model used by mathematicians as it leads to a clean partial differential equation for the displacement field.

A concise history of the development of the equations of the linearized theory of elasticity can be found in Love [24] and a detailed history of the subject can be found in Todhunter and Pearson [53, 54].

3.2 *Classical Nonlinear Theories of Elasticity*

In a Cauchy elastic body (Cauchy [9, 10]), the Cauchy stress is a function of the deformation gradient $\mathbf{F}$, and the density ρ, i.e.,

$$\mathbf{T} = \mathbf{f}(\rho, \mathbf{F}). \tag{15}$$

Thus the Cauchy stress depends only on the extent of deformation between the reference and current configuration and does not depend on the loading history. A Green elastic body (Green [16, 17]) is a special subclass of Cauchy elastic bodies wherein one assumes the existence of a stored energy function W that depends on the density and the deformation gradient, the stress being derivable from the stored energy. A detailed discussion of the basic issues concerning nonlinear elasticity can be found in Truesdell and Noll [58]. Some recent developments can be found in Rajagopal [37, 38].

4 Implicit Theories for Elastic Solids

Using standard tools of continuum mechanics, it can be shown that the Cauchy stress in the most general compressible isotropic Cauchy elastic solid has the representation

$$\mathbf{T} = \gamma_0 \mathbf{1} + \gamma_1 \mathbf{B} + \gamma_2 \mathbf{B}^2, \tag{16}$$

where the material moduli $\gamma_i, i = 0, 1, 2$ depend on ρ, $tr\mathbf{B}$, $tr\mathbf{B}^2$, $tr\mathbf{B}^3$, or the density and any set of mutually independent invariants of $\mathbf{B}$. The important take away from the representation is that the material moduli cannot depend on any invariants of the stress; a result clearly contradicted by experiments that unequivocally show that the material moduli do depend on the mean normal stress.

Let us change the starting point for the development of constitutive relations for elastic bodies and suppose that the density, the Cauchy stress and the deformation gradient are related through an implicit constitutive relation:

$$\mathbf{f}(\rho, \mathbf{T}, \mathbf{F}) = \mathbf{0}. \tag{17}$$

In the case of an isotropic body, it can be shown that the above relationship reduces to

$$\mathbf{f}(\rho, \mathbf{T}, \mathbf{B}) = \mathbf{0}. \tag{18}$$

The function $\mathbf{f}$ needs to obey

$$\mathbf{f}(\rho, \mathbf{QTQ}^T, \mathbf{QBQ}^T) = \mathbf{Qf}(\rho, \mathbf{T}, \mathbf{B})\mathbf{Q}^T, \quad \forall \mathbf{Q} \in O, \tag{19}$$

where O denotes the orthogonal group. It then follows that $\mathbf{f}$ has the following representation (see Spencer [52]):

$$\begin{aligned} &\alpha_0 \mathbf{1} + \alpha_1 \mathbf{T} + \alpha_2 \mathbf{B} + \alpha_3 \mathbf{T}^2 + \alpha_4 \mathbf{B}^2 + \alpha_5 (\mathbf{TB} + \mathbf{BT}) \\ &\quad + \alpha_6 (\mathbf{T}^2\mathbf{B} + \mathbf{BT}^2) + \alpha_7 (\mathbf{B}^2\mathbf{T} + \mathbf{TB}^2) + \alpha_8 (\mathbf{T}^2\mathbf{B}^2 + \mathbf{B}^2\mathbf{T}^2) = \mathbf{0} \end{aligned} \tag{20}$$

where the material moduli $\alpha_i, i = 0, \ldots, 8$ depend upon

$$\rho,\ tr\mathbf{T},\ tr\mathbf{B},\ tr\mathbf{T}^2,\ tr\mathbf{B}^2,\ tr\mathbf{T}^3,\ tr\mathbf{B}^3,\ tr(\mathbf{TB}),\ tr(\mathbf{T}^2\mathbf{B}),\ tr(\mathbf{B}^2\mathbf{T}),\ tr(\mathbf{T}^2\mathbf{B}^2). \tag{21}$$

Equation (16) defines a subclass of the general implicit model defined through (20) and (21). Notice a model of the form

$$\mathbf{B} = \tilde{\alpha}_0 \mathbf{1} + \tilde{\alpha}_1 \mathbf{T} + \tilde{\alpha}_2 \mathbf{T}^2, \tag{22}$$

where $\tilde{\alpha}_i, i = 0, 1, 2$ depending upon ρ, $tr\mathbf{T}$, $tr\mathbf{T}^2$, $tr\mathbf{T}^3$, is also a subclass of (20), (21).

Some effort has been expended in investigating the semi-invertibility and invertibility of isotropic tensors (see Truesdell and Moon [57]), i.e., given a tensor valued function $\mathbf{M} = \mathbf{f}(\mathbf{N})$ where $\mathbf{f}$ is an isotropic function, one can ask when one could express $\mathbf{N}$ as $\mathbf{N} = \mathbf{g}(\mathbf{M})$. It is interesting to note that the simple one-dimensional response of a spring and an inextensible string shown in Figs. 3, 4 does not allow for invertibility. Notice that the above response is non-dissipative and any energy stored

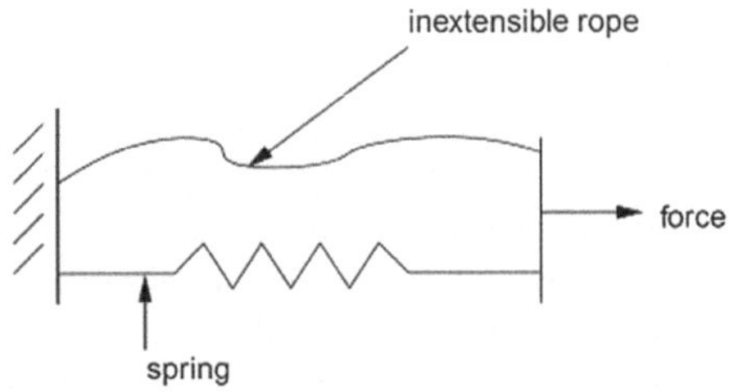

Fig. 3 Spring-inextensible string arrangement

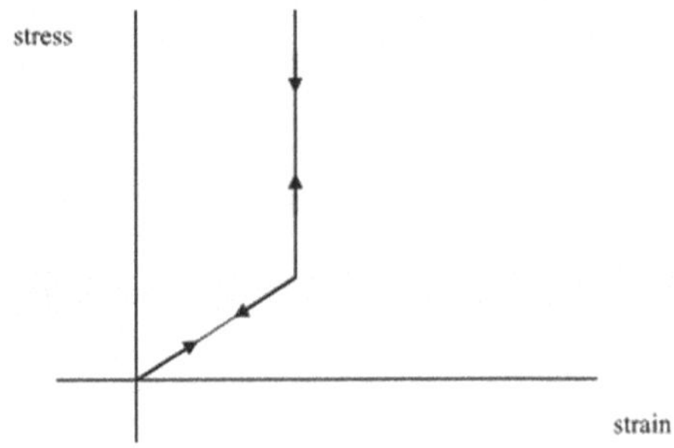

Fig. 4 Non-invertible response of the spring-string system expressed as a stress–strain relation (for an analogous elastic body) rather than a force–displacement relation

in the spring can be completely recovered. In its generalization to an elastic body, while the strain can be expressed as a function of stress, the converse is not possible (see Rajagopal [36] for a detailed discussion of models wherein the stress cannot be expressed as a function of the strain).

In a series of papers Rajagopal and co-workers have studied implicit constitutive relations for elasticity (Rajagopal [35, 37], Rajagopal and Srinivasa [44, 45], Rajagopal and Saccomandi [42]). Rajagopal and Srinivasa [45] provide a thermodynamic basis for the implicit theory of elasticity.

There are several advantages to using implicit models for describing the elastic response of solids: it is in keeping with the demands of causality, a fundamental underpinning of Newtonian physics. Unlike Cauchy elasticity, the implicit theory allows for the material moduli of the constitutive relation describing a body to depend on the mean normal stress as well as the other stress invariants and mixed invariants involving the stress and other kinematical quantities. It allows one to obtain in a rational manner a nonlinear relationship between the linearized strain and the stress that allows for a consistent approach to problems involving singularities in strains within the context of the linearized approximation for problems such as fracture and strains adjacent to the apex of notches. (Rajagopal and Walton [48], Gou et al. [15], Kulvait et al. [22], Zappalorto et al. [61], Itou et al. [19–21], Rajagopal and Zappalorto [49]). The implicit models permit one to describe the response of elastic bodies such as those that cannot be described within the context of classical Cauchy theory of elasticity (Rajagopal [40], Devendiran et al. [11]). The implicit elastic model is able to describe the response of several metallic alloys which exhibit nonlinearity even in the range of small strains where the linearized strain would be applicable. The classical linearized model represented by (13) or (14) is incapable of describing such phenomena.

Under the assumption (10), it follows from (22) that

$$\boldsymbol{\varepsilon} = \lambda_1 \mathbf{1} + \lambda_2 \mathbf{T} + \lambda_3 \mathbf{T}^2, \tag{23}$$

where the $\lambda_i, i = 1, 2, 3$ depend linearly on $tr\boldsymbol{\varepsilon}$, but arbitrarily on ρ and the invariants of $\mathbf{T}$. If we linearize (20) under the assumption (10), we find that it can be written as (on tacitly assuming that the $\hat{\alpha}_i$ as a function $\boldsymbol{\varepsilon}$ are not sub-linear)

$$\boldsymbol{\varepsilon} + \hat{\alpha}_1 \mathbf{1} + \hat{\alpha}_2 \mathbf{T} + \hat{\alpha}_3 \mathbf{T}^2 + \hat{\alpha}_4 (\mathbf{T}\boldsymbol{\varepsilon} + \boldsymbol{\varepsilon}\mathbf{T}) + \hat{\alpha}_5 (\mathbf{T}^2 \boldsymbol{\varepsilon} + \boldsymbol{\varepsilon}\mathbf{T}^2) = \mathbf{0}, \tag{24}$$

where the function $\hat{\alpha}_i, i = 1, 2, 3$ are scalar valued functions that can at most depend linearly on $tr\boldsymbol{\varepsilon}$, but arbitrarily on ρ and the invariants of $\mathbf{T}$ while $\hat{\alpha}_4$ and $\hat{\alpha}_5$ are functions of ρ, $tr\mathbf{T}$, $tr\mathbf{T}^2$ and $tr\mathbf{T}^3$. The expression (24) would only be valid if the coefficients $\alpha_i, i = 0, \ldots, 8$ and $\boldsymbol{\varepsilon}$ and $\mathbf{T}$ are such that each of the terms that appear in (24) are of $O(\delta)$. We shall assume that such is the case. The above approximation while requiring that the strain be small places no restrictions on the stress being small.

Within the context of (23), in the case of one-dimensional response of the body, we have the relationship

$$\varepsilon = f(\sigma) \tag{25}$$

where ε and σ are the one-dimensional strain and stress, respectively. It would be perfectly reasonable within the context of the above approach to have a constitutive model of the form

$$\varepsilon = K\sigma^m, \tag{26}$$

where K and m are constants. It immediately follows that we can express the stress in terms of the strain as

$$\sigma = E\varepsilon^n, \tag{27}$$

where $E = 1/K$ and $n = 1/m$. We have thus derived a constitutive relation that is used often in mechanics to describe the nonlinear behavior of solids (especially the inelastic response) at infinitesimal strains. Such an approximation would not be consistent if we were to derive it within the context of an explicit model wherein the stress is expressed explicitly in terms of the stretch tensor $\mathbf{B}$ as we can only obtain a linear relationship between the stress and the linearized strain were we to carry out the linearization by appealing to (10). However, when we have an implicit constitutive model of the form (20), we can without any inconsistency arrive at a model such as (27). For any nonlinear function $f(\sigma)$ such models are completely consistent with respect to our approximation procedure, as no restrictions are placed on the stress However, we have to ensure that both the left and right hand side of (25) are the same order of magnitude.

Most models used to describe the inelastic response of rigid bodies use implicit constitutive relations. We shall not discuss such relations here.

5 Implicit Models for Fluids

We now turn our attention to implicit models for fluids, but as before we will discuss briefly the explicit classical models that have been and are very popular. In fact, no model has enjoyed the kind of attention as that of the Navier–Stokes fluid. We start with the documentation of the Euler fluid. The Euler fluid is defined by a constitutive expression for the stress in terms of the density, namely,

$$\mathbf{T} = -p(\rho)\mathbf{1}, \tag{28}$$

where ρ is the density and p is referred to as the pressure. Euler [12–14] developed the above model in a series of papers, the first paper had an error that he rectified in the paper that followed. It is interesting to note that the fluid in fact belongs to the class of elastic bodies defined in the previous section. The body described by (28) is incapable of dissipating energy and is hence an elastic body.

The simplest classical model for a fluid that is capable of dissipating energy is the Navier–Stokes model. Navier [28] developed a model that had only one material modulus that defined the fluid. Later, Poisson [33] developed a model that has both the moduli that appear in the classical Navier–Stokes model given by the constitutive expression (29) below. Both Navier and Poisson appealed to molecular arguments. Stokes [50] re-derived the model of Poisson's, but from a phenomenological continuum perspective.

The stress in a compressible classical Navier–Poisson–Stokes model is usually expressed as

$$\mathbf{T} = -p(\rho)\mathbf{1} + \lambda(\rho)(tr\mathbf{D})\mathbf{1} + 2\mu(\rho)\mathbf{D}, \tag{29}$$

where

$$\mathbf{D} = \frac{1}{2}\left[\left(\frac{\partial \mathbf{v}}{\partial \mathbf{x}}\right) + \left(\frac{\partial \mathbf{v}}{\partial \mathbf{x}}\right)^T\right] \tag{30}$$

is the symmetric part of the velocity gradient and $\mathbf{v}$ is the velocity. While the material modulus μ is the shear viscosity and has physical significance, the modulus λ has no clear physical underpinning.

If we start, as is customary, with the constitutive assumption that

$$\mathbf{T} = \mathbf{f}(\rho, \mathbf{L}), \tag{31}$$

then frame indifference implies that

$$\mathbf{T} = \mathbf{f}(\rho, \mathbf{D}). \tag{32}$$

If we are interested in an isotropic fluid, then it follows that

$$\mathbf{T} = \alpha_0 \mathbf{1} + \alpha_1 \mathbf{D} + \alpha_2 \mathbf{D}^2, \tag{33}$$

where the α_i, $i = 0, 1, 2$ depends on the density ρ, invariants $tr\mathbf{D}$, $tr\mathbf{D}^2$ and $tr\mathbf{D}^3$ or any other set of mutually independent invariants. Demanding that the stress be linear in $\mathbf{D}$ leads to the classical compressible Navier–Stokes fluid model (29).

If one requires that the model (29) meets the constraint of incompressibility, then the incompressible Navier–Stokes model takes the form

$$\mathbf{T} = -p\mathbf{1} + 2\mu \mathbf{D}, \tag{34}$$

where p is the indeterminate scalar that enforces the constraint of incompressibility and the viscosity μ is a constant.

Suppose instead of (31) as the starting point, we were to start with the assumption that

$$\mathbf{L} = \mathbf{g}(\rho, \mathbf{T}). \tag{35}$$

Then just appealing to the balance of angular momentum and Galilean Invariance will lead to the symmetric part of the velocity gradient D (bold face) depending on the density and the Cauchy stress. (see Rajagopal and Srinivasa [46]). Then the requirement of isotropy and representation theorems will lead to

$$\mathbf{D} = \gamma_1 \mathbf{1} + \gamma_2 \mathbf{T} + \gamma_3 \mathbf{T}^2, \tag{36}$$

where γ_i, $i = 1, 2, 3$ depend on the principal invariants of the Cauchy stress and the density ρ. Demanding that the model (36) be linear in $\mathbf{T}$ then once again leads to the classical Navier–Stokes model, but with an expression for the symmetric part of the velocity gradient in terms of the stress.

If instead of expressing the stress $\mathbf{T}$ as a function of $\mathbf{D}$, we express $\mathbf{D}$ as a function of $\mathbf{T}$, (29) takes the form

$$\mathbf{D} = \alpha \left[\mathbf{T} - \frac{1}{3}(tr\mathbf{T})\mathbf{1} \right] + \beta \left[p(\rho) + \frac{1}{3}(tr\mathbf{T}) \right] \mathbf{1}, \tag{37}$$

where

$$\alpha = \frac{1}{2\mu} \quad \text{and} \quad \beta = \frac{1}{3\lambda + 2\mu}, \tag{38}$$

wherein both μ and $(3\lambda + 2\mu)$, and hence α and β, have clear physical meaning. The term (1/3) $(3\lambda + 2\mu)$ is the bulk modulus of the fluid that can be measured easily by means of an experiment. We also notice that the representation (37) very elegantly splits the symmetric part of the velocity gradient into a deviatoric part and a spherical part.

When the fluid is incompressible, as it can undergo only isochoric motion we find that from (37) that

$$p(\rho) = -\frac{1}{3}(tr\mathbf{T}), \tag{39}$$

the density of course being constant at a specific material point as the fluid is incompressible (the density can vary from one material point to another if the fluid is inhomogeneous), and hence the thermodynamic pressure is the mean normal stress.

Stokes suggested that for a class of flows it might be reasonable to assume that $(3\lambda + 2\mu) = 0$. Unfortunately, the above suggestion which goes by the name "Stokes Conjecture" has been used indiscriminately. Despite Stokes [51] himself having serious second thoughts about the assumption we find it used frequently (A detailed discussion concerning the inaptness of the assumption can be found in Rajagopal [39]). We notice from (37) and (38) that the one would never even think along the lines of the "Stokes Conjecture" were one to start the development of the model from (35), which would indeed be the correct approach that meets the requirements of causality. The "Stokes Conjecture" is not a tenable assumption as for any finite value of β the sum of $(3\lambda + 2\mu)$ cannot be zero.

The main consequence of causality is to recognize that the starting point for describing linear constitutive relations for compressible fluids ought to be the constitutive relation

$$\mathbf{D} = \beta_1(\rho)\mathbf{1} + \beta_2(\rho)(tr\mathbf{T})\mathbf{1} + \beta_3(\rho)\mathbf{T}. \tag{40}$$

Rajagopal [39] obtains a relationship between the material coefficients that appear in (37) and (40), namely,

$$\beta_1 = \frac{p}{3\lambda + 2\mu}, \quad \beta_2 = \frac{-\lambda}{2\mu(3\lambda + 2\mu)}, \quad \beta_3 = \frac{1}{2\mu}. \tag{41}$$

We now turn our attention to generalizations of the classical Navier–Stokes model. While the responses (29) and (30) might be useful in describing the response of a large class of fluids, it becomes necessary to use a fully implicit relation to describe an incompressible fluid with a viscosity that depends on the mean normal stress and the shear rate (see Hron et al. [18]). In fact, such models are used to describe the response of a large class of lubricants and organic liquids. Consider a constitutive relation expressed in the form

$$\mathbf{T} = -p\mathbf{1} + \mu(tr\mathbf{T}, tr\mathbf{D}^2)\mathbf{D}. \tag{42}$$

The above expression is not in general an explicit expression for the stress in terms of the symmetric part of the velocity gradient but is an implicit relationship that belongs to the general class defined through

$$\mathbf{f}(\rho, \mathbf{T}, \mathbf{D}) = \mathbf{0}. \tag{43}$$

Let us consider the implications of assuming that **f** defined through the relation (43) is an isotropic function of the tensors **T** and **D**. Then

$$\mathbf{Q}\mathbf{f}(\rho, \mathbf{T}, \mathbf{D})\mathbf{Q}^T = \mathbf{f}(\rho, \mathbf{Q}\mathbf{T}\mathbf{Q}^T, \mathbf{Q}\mathbf{D}\mathbf{Q}^T), \quad \forall \mathbf{Q} \in O, \tag{44}$$

where O denotes the set of all orthogonal transformations. It then follows that (see Spencer [52]):

$$\begin{aligned} &\alpha_0\mathbf{1} + \alpha_1\mathbf{T} + \alpha_2\mathbf{D} + \alpha_3\mathbf{T}^2 + \alpha_4\mathbf{D}^2 + \alpha_5(\mathbf{D}\mathbf{T} + \mathbf{T}\mathbf{D}) \\ &\quad + \alpha_6(\mathbf{T}^2\mathbf{D} + \mathbf{D}\mathbf{T}^2) + \alpha_7(\mathbf{D}^2\mathbf{T} + \mathbf{T}\mathbf{D}^2) + \alpha_8(\mathbf{T}^2\mathbf{D}^2 + \mathbf{D}^2\mathbf{T}^2) = \mathbf{0} \end{aligned} \tag{45}$$

where the material functions $\alpha_i, i = 0, \ldots, 8$ depend on the density ρ and the invariants

$$tr\mathbf{T},\ tr\mathbf{D},\ tr\mathbf{T}^2,\ tr\mathbf{D}^2,\ tr\mathbf{T}^3,\ tr\mathbf{D}^3,\ tr(\mathbf{T}\mathbf{D}),\ tr(\mathbf{T}^2\mathbf{D}),\ tr(\mathbf{D}^2\mathbf{T}),\ tr(\mathbf{T}^2\mathbf{D}^2). \tag{46}$$

The model which allows the viscosity to depend on the mean normal stress and the shear rate, namely,

$$\mathbf{T} = \frac{1}{3}(tr\mathbf{T})\mathbf{1} + 2\left[\mu(\mathbf{T}, \mathbf{D})\right]\mathbf{D}. \tag{47}$$

considered by Hron et al. [18] is a special subclass of (45) and (46). This is a truly implicit model in general in that neither **T** can be expressed explicitly in terms of **D** or vice-versa. Once again, in the above model, when $\mathbf{D} = \mathbf{0}$, the stress is purely spherical.

The classical compressible and incompressible power-law fluid are described by the following constitutive expressions, respectively,

$$\mathbf{T} = -p(\rho)\mathbf{1} + \mu(\rho)\left[1 + \delta(tr\mathbf{D}^2)\right]^n \mathbf{D}, \tag{48}$$

and

$$\mathbf{T} = -p\mathbf{1} + \mu\left[1 + \delta(tr\mathbf{D}^2)\right]^n \mathbf{D}.$$

The counterpart to the generalized compressible power-law fluid takes the following form:

$$\mathbf{D} = \alpha\left\{1 + \gamma(tr\mathbf{T}^2)^{\frac{1}{2}}\right\}^n \mathbf{T}, \tag{49}$$

where α and γ could be depend on the density. If the fluid is incompressible, we would require that $tr\mathbf{D} = \mathbf{0}$. The model (see Malek et al. [25])

$$\mathbf{D} = \alpha\left\{1 + \gamma\left[tr\left(\mathbf{T} - \frac{1}{3}(tr\mathbf{T})\mathbf{1}\right)\right]^2\right\}^n\left[\mathbf{T} - \frac{1}{3}(tr\mathbf{T})\mathbf{1}\right], \tag{50}$$

automatically satisfies the constraint of incompressibility. The counterpart of the generalized incompressible Stokesian fluid takes the form:

$$\mathbf{D} = \alpha_1\left[\mathbf{T} - \frac{1}{3}(tr\mathbf{T})\mathbf{1}\right] + \alpha_2\left[\mathbf{T}^2 - \frac{1}{3}(tr\mathbf{T}^2)\mathbf{1}\right]. \tag{51}$$

It should be clear from the discussion above that one could start with the relation (43) and proceed along the same lines as that considered for the incompressible case to obtain the constitutive relations for compressible fluids.

Recently, Perlacova and Prusa [31] have considered models belonging to sub-classes of (45) to describe experimental results concerning colloids and suspensions which are not possible within the ambit of models of the class (29). Le Roux and Rajagopal [23] used the constitutive expression:

$$\mathbf{D} = \alpha\left\{\left[1 + \beta(tr\mathbf{T}_d^2)\right]^n + \gamma\right\}\mathbf{T}_d, \tag{52}$$

where α, β, γ and n are constants to describe the non-monotonic response exhibited by many colloids and suspensions, and $\mathbf{T}_d$ denotes the deviatoric part of $\mathbf{T}$.

Several simple flows of a fluid modeled by (52) have been studied recently (see Malek et al. [25], Perlacova and Prusa [31], Narayan and Rajagopal [27], Rajagopal, Saccomandi and Vergori [43], and Gomez-Constante and Rajagopal (2018)).

With regard to describing the response of viscoelastic fluids that exhibit stress relaxation, most of the models that are used to describe such fluids are implicit models. While the classical Maxwell model (Maxwell [26]) is not an implicit model in that it expresses the symmetric part of the velocity gradient in terms of the stress and an appropriate time derivative of the stress, the models developed by Oldroyd [30] and Burgers [4] are truly implicit models relating the symmetric part of the velocity gradient and its proper frame-indifferent time derivative and the stress and its proper frame-indifferent time derivatives. We shall not discuss such models here.

6 Concluding Remarks

The response of many real bodies cannot be described by constitutive expression for the stress in terms of kinematical variables. Many responses require implicit relationship between the stress and its frame-indifferent material time derivatives and kinematical quantities and their frame-indifferent material time derivatives. Also, from the philosophical standpoint, providing expressions for the stress in terms of kinematical quantities is upside-down as it is contrary to the demands of causality. However, at times we are not in a position to clearly identify the causes and the effects, especially when very many quantities that are relevant to the response are being considered thus requiring implicit constitutive relations. In this short paper, we have developed algebraic constitutive relations for describing the response of elastic bodies and also fluids that have material properties that depend on the invariants of the stress.

Acknowledgements Rajagopal thanks the Office of Naval Research for its support of the work.

References

1. Boltenhagen, P., Hu, Y., Matthys, E., Pine, D.: Observation of bulk phase separation and coexistence in a sheared micellar solution. Phys. Rev. Lett. **79**, 2359–2362 (1997)
2. Bulicek, M., Malek, J., Rajagopal, K.R., Suli, E.: On elastic solids with limiting small strain: modelling and analysis. EMS Surv. Math. Sci. **1**, 283–332 (2014)
3. Bulicek, M., Malek, J., Rajagopal, K.R., Walton, J.R.: Existence of solutions for the anti-plane stress for a new class of "strain-limiting" elastic bodies. Calcul. Var. Part. Differ. Equ. **54**, 2115–2147 (2015)
4. Burgers, J.M.: Mechanical considerations-model systems-phenomenological theories of relaxation and of viscosity. In: First Report on Viscosity and Plasticity, 3nd edn. Nordemann Publishing Company, Inc., New York (1939)
5. Bustamante, R., Rajagopal, K.R.: Solutions of some simple boundary value problems within the context of a new class of elastic materials. Int. J. Non-linear Mech. **46**, 376–386 (2011)
6. Bustamante, R., Rajagopal, K.R.: On a new class of electoro-elastic bodies. I. Proc. R. Soc. Lond. Ser. A **469**, 20120521 (2013)
7. Bustamante, R., Rajagopal, K.R.: On a new class of electro-elastic bodies. II. Boundary value problems. Proc. R. Soc. Lond. Ser. A **469**, 20130106 (2013)
8. Bustamante, R., Rajagopal, K.R.: Implicit constitutive equations for nonlinear magnetoelastic bodies. Proc. R. Soc. Lond. Ser. A **471**, 20140959 (2015)
9. Cauchy, A.L.: Recherches sur lequilibre et le mouvement interieur des corps solides ou fluids, elastiques ou non elastiques. Bull. Soc. Philomath 9–13 (1823); Oeuvres (Series II) **2**, 300–304 (1823)
10. Cauchy, A.L.: Sur les equations qui experiments des conditions de equilibre ou le lois du mouvement interieur, d'un corps solide, elastique, ou non elastique. Ex. de Mathematique **3**, 160–187 (1828); Oeuvres (Series II) **8**, 195–226 (1828)
11. Devendiran, V.K., Sandeep, R.K., Kannan, K., Rajagopal, K.R.: A thermodynamically consistent constitutive equation for describing the response exhibited by several alloys and the study of a meaningful physical problem. Int. J. Solids Struct. **108**, 1–10 (2017)
12. Euler, L.: Sur le mouvement de l'eau par des tuyaux de conduite. Memoires de l'academie des sciences de Berlin **8**, 111–148 (1754); Opera Omnia, Series 2, **15**, 219–250, E206

13. Euler, L.: Principes generaux du mouvement des fluids. Memoires de l'academie des sciences de Berlin **11**, 274–315 (1757); Opera Omnia, Series 2, **12**, 54–91, E226
14. Euler, L.: Principia motus fluidorum. in Novi Commentarii academiae scientiarum Petropolitanae **6**, 271–311 (1761); Opera Omnia, Series 2, **12**, 133–168, E258
15. Gou, K., Mallikarjuna, M., Rajagopal, K.R., Walton, J.R.: Modeling fracture in the context of a strain-limiting theory of elasticity: a single plane-strain crack. Int. J. Eng. Sci. **88**, 73–82 (2015)
16. Green, G.: On the laws of reflection and refraction of light at the common surface of two non-crystallized media. Trans. Cambridge Phil. Soc. **7**, 1–24 (1838–1842); Papers 245–269 (1839)
17. Green, G.: On the propagation of light in crystallized media. Trans. Cambridge Phil. Soc. **7**, 121–140 (1838–1842); Papers 293–311 (1841)
18. Hron, J., Malek, J., Rajagopal, K.R.: Simple flows of fluids with pressure-dependent viscosities. Proc. R. Soc. Lond. Ser. A **457**, 1603–1622 (2001)
19. Itou, H., Kovtunenko, V.A., Rajagopal, K.R.: Nonlinear elasticity with limiting small strain for cracks subject to non-penetration. Math. Mech. Solids **22**, 1334–1346 (2017)
20. Itou, H., Kovtunenko, V.A., Rajagopal, K.R.: Contacting crack faces within the context of bodies exhibiting limiting strains. JSIAM Lett. **9**, 61–64 (2017)
21. Itou, H., Kovtunenko, V.A., Rajagopal, K.R.: On the states of stress and strain adjacent to a crack in a strain-limiting viscoelastic body. Math. Mech. Solids **23**, 433–444 (2018)
22. Kulvait, V., Malek, J., Rajagopal, K.R.: Anti-plane stress state of a plate with a V-notch for a new class of elastic solids. Int. J. Fracture **179**, 59–73 (2013)
23. Le Roux, C., Rajagopal, K.R.: Shear flows of a new class of power-law fluids. Appl. Math. **58**, 153–177 (2013)
24. Love, A.E.H.: A Treatise on the Mathematical Theory of Elasticity. Dover Publications, New York (1944)
25. Malek, J., Prusa, V., Rajagopal, K.R.: Generalizations of the Navier-Stokes fluid from a new perspective. Int. J. Eng. Sci. **48**, 1907–1924 (2010)
26. Maxwell, J.C.: On the dynamical theory of gases. Phil. Trans. R. Soc. Lond. **157**, 49–88 (1867)
27. Narayan, S.A., Rajagopal, K.R.: Unsteady flows of a class of novel generalizations of the Navier-Stokes fluid. Appl. Math. Comput. **219**, 9935–9946 (2013)
28. Navier, C.L.M.H.: Sur les lois de l'equilibre et du mouvement des corps solides elastiques. Bull. Soc. Philomath. **177–181** (1821)
29. Newton, I.: Philosophiae naturalis principia mathematica. (1687), (translated by Andrew Motte) Prometheus Books, New York (1995)
30. Oldroyd, J.G.: On the formulation of rheological equations of state. Proc. R. Soc. Lond. Ser. A **200**, 523–541 (1950)
31. Perlacova, T., Prusa, V.: Tensorial implicit constitutive relations in mechanics of incompressible Non-Newtonian fluids. J. Non-Newtonian Fluid Mech. **216**, 13–21 (2015)
32. Poisson, S.D.: Memoire sur l'equilibre et le mouvement des corps elastiques. Mem. Acad. Sci. Inst. France **8**, 357–570 (1829)
33. Poisson, S.D.: Memoire sur les equations generales de l'equilibre et du mouvement des corps solides elastiques et des fluides. J. de l'Ecole Polytechnique **13**, 1–174 (1831)
34. Prusa, V., Rajagopal, K.R.: On implicit constitutive relations for materials with fading memory. J. Non-Newtonian Fluid Mech. **181–182**, 22–29 (2012)
35. Rajagopal, K.R.: On implicit constitutive theories. Appl. Math. **48**, 279–319 (2003)
36. Rajagopal, K.R.: On implicit constitutive theories for fluids. J. Fluid Mech. **550**, 243–249 (2006)
37. Rajagopal, K.R.: The elasticity of elasticity. Z. Angew. Math. Phys. **58**, 309–317 (2007)
38. Rajagopal, K.R.: Conspectus of concepts of elasticity. Math. Mech. Solids **16**, 536–562 (2011)
39. Rajagopal, K.R.: A new development and interpretation of the Navier-Stokes fluid which reveals why the "Stokes assumption" is inapt. Int. J. Non-linear Mech. **50**, 141–151 (2013)
40. Rajagopal, K.R.: On the nonlinear elastic response of bodies in the small strain range. Acta Mech. **225**, 1545–1553 (2014)

41. Rajagopal, K.R.: A note on the classification of anisotropy of bodies defined by implicit constitutive relations. Mech. Res. Commun. **64**, 38–41 (2015)
42. Rajagopal, K.R., Saccomandi, G.: The mechanics and mathematics of the effect of pressure on the shear modulus of elastomers. Proc. R. Soc. Lond. Ser. A **465**, 3859–3874 (2009)
43. Rajagopal, K.R., Saccomandi, G., Vergori, L.: Flow of fluids with pressure- and shear-dependent viscosity down an inclined plane. J. Fluid Mech. **706**, 173–189 (2012)
44. Rajagopal, K.R., Srinivasa, A.R.: On the response of non-dissipative solids. Proc. R. Soc. Lond. Ser. A **463**, 357–367 (2007)
45. Rajagopal, K.R., Srinivasa, A.R.: On a class of non-dissipative materials that are not hyperelastic. Proc. R. Soc. Lond. Ser. A **465**, 493–500 (2009)
46. Rajagopal, K.R., Srinivasa, A.R.: Restrictions placed on constitutive relations by angular momentum balance and Galilean invariance. Z. Angew. Math. Phys. **64**, 391–401 (2013)
47. Rajagopal, K.R., Tao, L.: On the response of non-dissipative solids. Commun. Nonlinear Sci. Numer. Simul. **13**, 1089–1100 (2008)
48. Rajagopal, K.R., Walton, J.R.: Modeling fracture in the context of a strain-limiting theory of elasticity: a single anti-plane shear crack. Int. J. Fracture **169**, 39–48 (2011)
49. Rajagopal, K.R., Zappalorto, M.: Bodies described by non-monotonic strain-stress constitutive equations containing a crack subject to anti-plane shear stress. Int. J. Mech. Sci. **149**, 494–499 (2018)
50. Stokes, G.G.: On the theories of the internal friction of fluids in motion, and of the equilibrium and motion of elastic solids. Trans. Cambridge Philos. Soc. **8**, 287–319 (1845)
51. Stokes, G.G.: On the effect of the internal friction of fluids on the motion of pendulums. Trans. Cambridge Philos. Soc. **9**, 8–106 (1851)
52. Spencer, A.J.M.: Theory of invariants. In: Eringen, A.C. (ed.) Continuum physics I, (Part III), pp. 239–353. Academic Press, New York (1971)
53. Todhunter, I.: A History of the Theory of Elasticity and of the Strength of Materials, Volume I: From Galilei to St. Venant, edited and completed by K. Pearson. Cambridge University Press, Cambridge (1886)
54. Todhunter, I.: A History of the Theory of Elasticity and of the Strength of Materials, Volume II: From Saint-Venant to Lord Kelvin, edited and completed by K. Pearson. Cambridge University Press, Cambridge (1893)
55. Truesdell, C.: The Elements of Continuum Mechanics. Springer, Berlin (1966)
56. Truesdell, C.: A First Course in Rational Continuum Mechanics. Academic Press, London (1977)
57. Truesdell, C., Moon, H.: Inequalities sufficient to ensure semi-invertibility of isotropic functions. J. Elasticity **5**, 183–189 (1975)
58. Truesdell, C., Noll, W.: The Non-Linear Field Theories of Mechanics, 2nd edn. Springer, Berlin (1992)
59. Young, T.: A Syllabus of a Course of Lectures on Natural and Experimental Philosophy. The Press of the Royal Institution, W. Savage, Printer, London (1802)
60. Young, T.: A Course of Lectures on Natural Philosophy and the Mechanical Arts, In two volumes. St. Paul's Church Yard, by W. Savage, Printer, London, Printed for Joseph Johnson (1807)
61. Zappalorto, M., Berto, F., Rajagopal, K.R.: On the anti-plane state of stress near pointed or sharply radiused notches in strain limiting elastic materials: closed form solution and implications for fracture assessements. Int. J. Fracture **199**, 169–184 (2016)
62. J. Gomez-Constante and K. R. Rajagopal, Flow of a new class of non-Newtonian fluids in tubes of non-circular cross-sections, Phil. Trans. R. Society, A 377, 20180069 (2019)

On the Perturbation of Bleustein–Gulyaev Waves in Piezoelectric Media

Gen Nakamura, Kazumi Tanuma, and Xiang Xu

Abstract The Bleustein–Gulyaev (BG) waves are subsonic surface waves which propagate along the surface of a piezoelectric half-space whose constituent material has a hexagonal symmetry, and which satisfies the mechanically free and electrically closed boundary condition. We give a perturbation to the material constants of the piezoelectric half-space of hexagonal symmetry, which consists of a perturbative part of the elasticity tensor, a perturbative part of the piezoelectric tensor and a perturbative part of the dielectric tensor. We will then present a first-order perturbation formula for the phase velocity of BG waves, which expresses the shift in the velocity from its comparative value for a hexagonal piezoelectric half-space, caused by those perturbative parts of the three tensors. It can be observed that only a few components of the perturbative parts of the tensors can affect the first-order perturbation of the phase velocity of BG waves.

1 Introduction

Piezoelectric materials have been used in many engineering devices because of their intrinsic direct and converse piezoelectric effects that take place between electric fields and mechanical deformations [11, 13, 17]. In the system of the constitutive equations, the mechanical stress and the electric displacement are related to the mechanical displacement and the electric potential through the elasticity tensor, the

G. Nakamura
Department of Mathematics, Graduate School of Science, Hokkaido University, Sapporo, Japan
e-mail: nakamuragenn@gmail.com

K. Tanuma (✉)
Department of Mathematics, Faculty of Science and Technology,
Gunma University, Kiryu, Japan
e-mail: tanuma@gunma-u.ac.jp

X. Xu
School of Mathematical Sciences, Zhejiang University, Hangzhou, People's Republic of China
e-mail: xxu@zju.edu.cn

H. Itou et al. (eds.), *Mathematical Analysis of Continuum Mechanics and Industrial Applications III*, Mathematics for Industry 34,
https://doi.org/10.1007/978-981-15-6062-0_5

piezoelectric tensor, and the dielectric tensor, and it is the piezoelectric tensor through which the elastic fields and electric fields can be coupled with each other.

Surface waves have been a topic of utmost importance in materials science, seismology, and nondestructive evaluation [4]. When the piezoelectric material has a hexagonal symmetry, which means that the material has one sixfold symmetry axis, a subsonic surface wave called the Bleustein–Gulyaev (BG) wave [2, 5] propagates along the surface of a piezoelectric half-space, provided that the sixfold symmetry axis lies on the surface and the propagation direction on the surface is perpendicular to the sixfold axis. Its phase velocity is written explicitly in terms of the aforementioned three material tensors.

Now we give a perturbation to the three material tensors of the piezoelectric half-space of hexagonal symmetry. This perturbation consists of a perturbative part of the elasticity tensor (which has 21 components), a perturbative part of the piezoelectric tensor (has 18 components), and a perturbative part of the dielectric tensor (has 6 components), for which we do not assume any material symmetry. We then observe how these perturbative tensors (which have 45 components in all) affect the phase velocity of BG waves that propagate along the surface of the piezoelectric half-space.

Thus, we will investigate the perturbation of the phase velocity of BG waves, which expresses the shift in the velocity from its comparative value for the piezoelectric half-space of hexagonal symmetry, caused by the aforementioned perturbative parts of the three tensors. We will then present a velocity formula which is correct to within terms linear in those perturbative parts of the tensors. The result is somewhat surprising; only a few components of the perturbative parts of the tensors can affect the first-order perturbation of the phase velocity of BG waves.

When the piezoelectric tensor vanishes, the system of the equations can be decomposed into the system of elasticity equations and the equation of dielectricity. In an anisotropic elastic half-space, the perturbation of phase velocity of subsonic surface waves (Rayleigh waves) has been well studied (see, e.g., [15, 16]). In comparison with the results therein, it can be observed that the first-order perturbation of the velocity of BG waves is affected by less number of the components of the perturbative part of the elasticity tensor.

Because of a limitation of the pages, we had to omit the details of the proof. Instead, we have included introductory descriptions and a motivation of this study so that this expository article becomes as understandable and as self-contained as possible. One objective of this article is to introduce the reader to the forthcoming full paper [12].

2 Preliminary

In the Cartesian coordinate system $\mathbf{x} = (x_1, x_2, x_3)$, the mechanical stress tensor $\boldsymbol{\sigma} = (\sigma_{ij})_{i,j=1,2,3}$ and the electric displacement $\boldsymbol{D} = (D_1, D_2, D_3)$ are related to the mechanical displacement $\boldsymbol{u} = (u_1, u_2, u_3)$ and the electric potential ϕ by the following constitutive equations:

$$\sigma_{ij} = \sum_{k,l=1}^{3} C_{ijkl} \frac{\partial u_k}{\partial x_l} + \sum_{l=1}^{3} e_{ijl} \frac{\partial \phi}{\partial x_l}, \qquad i, j = 1, 2, 3, \tag{1}$$

$$D_j = \sum_{k,l=1}^{3} e_{klj} \frac{\partial u_k}{\partial x_l} - \sum_{l=1}^{3} \varepsilon_{jl} \frac{\partial \phi}{\partial x_l}, \qquad j = 1, 2, 3.$$

Here $\mathbf{C} = (C_{ijkl})_{i,j,k,l=1,2,3}$ is the elasticity tensor, $\boldsymbol{e} = (e_{ijl})_{i,j,l=1,2,3}$ is the piezoelectric tensor, and $\boldsymbol{\varepsilon} = (\varepsilon_{jl})_{j,l=1,2,3}$ is the dielectric tensor. Hence the elastic and electric fields are coupled through the piezoelectric tensor.

These three tensors $\mathbf{C}$, $\boldsymbol{e}$, and $\boldsymbol{\varepsilon}$ satisfy the following symmetry conditions:

$$C_{ijkl} = C_{jikl} = C_{klij}, \quad e_{ijl} = e_{jil}, \quad \varepsilon_{jl} = \varepsilon_{lj}, \qquad i, j, k, l = 1, 2, 3.$$

We assume that the internal energy function is positive, which implies the following positivity conditions for $\mathbf{C}$ and $\boldsymbol{e}$:

$$\sum_{i,j,k,l=1}^{3} C_{ijkl}\, s_{ij}\, s_{kl} > 0, \qquad \sum_{j,l=1}^{3} \varepsilon_{jl}\, E_j\, E_l > 0 \tag{2}$$

for any non-zero 3×3 real symmetric matrix (s_{ij}) and for any non-zero real vector $(E_1, E_2, E_3) \in \mathbb{R}^3$.

The equations of mechanical motion with zero body force and the equation of electric equilibrium with zero free charge are given by

$$\sum_{j=1}^{3} \frac{\partial \sigma_{ij}}{\partial x_j} = \rho \frac{\partial^2 u_i}{\partial t^2}, \quad i = 1, 2, 3 \quad \text{and} \quad \sum_{j=1}^{3} \frac{\partial D_j}{\partial x_j} = 0, \tag{3}$$

respectively, where ρ is the uniform density. Substituting (1) into the preceding equations leads us to a system of the equations for the unknowns $\boldsymbol{u} = (u_1, u_2, u_3)$ and ϕ.

Let $\mathbf{m} = (m_1, m_2, m_3)$ and $\mathbf{n} = (n_1, n_2, n_3)$ be orthogonal unit vectors in $\mathbb{R}^3$, and consider surface waves in the piezoelectric half-space $\mathbf{n} \cdot \mathbf{x} \leq 0$ which propagate along the surface $\mathbf{n} \cdot \mathbf{x} = 0$ in the direction of $\mathbf{m}$ and decay exponentially as $\mathbf{n} \cdot \mathbf{x} \longrightarrow -\infty$. The general form of the solutions to (3) which describe such surface waves is written as

$$\begin{pmatrix} \boldsymbol{u} \\ \phi \end{pmatrix} = \sum_{\alpha=1}^{4} c_\alpha\, \mathbf{a}_\alpha\, \mathrm{e}^{-\mathrm{i}\, k(\mathbf{m}\cdot\mathbf{x} + p_\alpha \mathbf{n}\cdot\mathbf{x} - v\, t)} \tag{4}$$

in $\mathbf{n} \cdot \mathbf{x} \leq 0$, where $\mathrm{i} = \sqrt{-1}$, k is the wave number, v is the phase velocity in the subsonic range, and $\mathbf{a}_\alpha \in \mathbb{C}^4$ $(1 \leq \alpha \leq 4)$ and $p_\alpha \in \mathbb{C}$ $(1 \leq \alpha \leq 4)$ with $\mathrm{Im}\, p_\alpha > 0$ (the imaginary part of p_α) are determined from the Eq. (3) and c_α $(1 \leq \alpha \leq 4)$ are arbitrary complex constants.

There are two typical examples of boundary conditions. One is the "mechanically free and electrically open condition", which means that the mechanical traction and the normal component of the electric displacement vanish at the boundary, namely,

$$\sum_{j=1}^{3} \sigma_{ij} n_j = 0 \ \ (i = 1, 2, 3) \ \ \text{and} \ \ \sum_{i=1}^{3} D_i n_i = 0 \ \ \text{at } \mathbf{n} \cdot \mathbf{x} = 0. \tag{5}$$

The other is the "mechanically free and electrically closed (or, grounded) condition", which means that the mechanical traction and the electric potential vanish at the boundary, namely,

$$\sum_{j=1}^{3} \sigma_{ij} n_j = 0 \ \ (i = 1, 2, 3) \ \ \text{and} \ \ \phi = 0 \ \ \text{at } \mathbf{n} \cdot \mathbf{x} = 0. \tag{6}$$

The complex constants c_α $(1 \leq \alpha \leq 4)$ in (4) are determined from (each of) the preceding boundary conditions.

The Bleustein–Gulyaev (BG) waves are subsonic surface waves which satisfy the latter boundary condition; they propagate along the surface of a piezoelectric half-space under the mechanically free and electrically closed boundary condition, provided that the medium has a hexagonal symmetry, its sixfold symmetry axis lies on that surface and that the propagation direction is perpendicular to the sixfold axis, the details of which we shall return to in the next section.[1]

The constitutive equations (1) can be rewritten in terms of a nine-dimensional system by introducing the contracted notation called the Voigt notation. For $C_{ijkl} \in \mathbf{C}$ we use the rules of replacing the subscript ij (or kl) by α (or β) as follows:

ij (or kl)		α or (β)	ij (or kl)		α or (β)
11	$\longleftrightarrow$	1	23 or 32	$\longleftrightarrow$	4
22	$\longleftrightarrow$	2	31 or 13	$\longleftrightarrow$	5
33	$\longleftrightarrow$	3	12 or 21	$\longleftrightarrow$	6.

In the similar way, we replace the subscript ij (the first two indices) of $e_{ijl} \in \boldsymbol{e}$ by α.[2] The constitutive equations (1) are then rewritten as

[1] For the analysis of elastic surface waves (Rayleigh waves) and BG waves under the inhomogeneous boundary conditions, we refer to [6, 7], where a slow time perturbation procedure is applied to derive asymptotic formulations for those waves.

[2] We do not use these rules of replacing for $\boldsymbol{\sigma} = (\sigma_{ij})_{i,j=1,2,3}$ and $\boldsymbol{\varepsilon} = (\varepsilon_{jl})_{j,l=1,2,3}$.

$$\begin{bmatrix} \sigma_{11} \\ \sigma_{22} \\ \sigma_{33} \\ \sigma_{23} \\ \sigma_{13} \\ \sigma_{12} \\ \hline D_1 \\ D_2 \\ D_3 \end{bmatrix} = \left[\begin{array}{cccccc|ccc} C_{11} & C_{12} & C_{13} & C_{14} & C_{15} & C_{16} & e_{11} & e_{12} & e_{13} \\ & C_{22} & C_{23} & C_{24} & C_{25} & C_{26} & e_{21} & e_{22} & e_{23} \\ & & C_{33} & C_{34} & C_{35} & C_{36} & e_{31} & e_{32} & e_{33} \\ & & & C_{44} & C_{45} & C_{46} & e_{41} & e_{42} & e_{43} \\ & & & & C_{55} & C_{56} & e_{51} & e_{52} & e_{53} \\ Sym. & & & & & C_{66} & e_{61} & e_{62} & e_{63} \\ \hline & & & & & & -\varepsilon_{11} & -\varepsilon_{12} & -\varepsilon_{13} \\ & & & & & & & -\varepsilon_{22} & -\varepsilon_{23} \\ & & & & & & & & -\varepsilon_{33} \end{array}\right] \begin{bmatrix} (u_1)_{x_1} \\ (u_2)_{x_2} \\ (u_3)_{x_3} \\ (u_2)_{x_3} + (u_3)_{x_2} \\ (u_1)_{x_3} + (u_3)_{x_1} \\ (u_1)_{x_2} + (u_2)_{x_1} \\ \hline \phi_{x_1} \\ \phi_{x_2} \\ \phi_{x_3} \end{bmatrix}$$

The 9×9 symmetric matrix which appears in the right-hand side of the preceding equation is called the elasto-piezo-dielectric matrix [13] and we denote it by P.

3 Bleustein–Gulyaev (BG) Waves

We assume that the piezoelectric medium has a hexagonal symmetry, which means that the medium has one sixfold symmetry axis, i.e., a symmetry axis of the rotation of degree $2\pi/6$. Let the 3-axis be the sixfold axis. Then the non-zero components of $\mathbf{C}$, $\boldsymbol{e}$ and $\boldsymbol{\varepsilon}$ are

$$C_{11}, C_{22}, C_{33}, C_{12}, C_{13}, C_{23}, C_{44}, C_{55}, C_{66}, \quad e_{13}, e_{23}, e_{33}, e_{41}, e_{42}, e_{51}, e_{52}, \quad \varepsilon_{11}, \varepsilon_{22}, \varepsilon_{33}$$

and they satisfy

$$C_{11} = C_{22}, \quad C_{13} = C_{23}, \quad C_{44} = C_{55}, \quad C_{66} = \frac{C_{11} - C_{22}}{2},$$
$$e_{13} = e_{23}, \quad e_{42} = e_{51}, \quad -e_{41} = e_{52}, \quad \varepsilon_{11} = \varepsilon_{22}.$$

The elasto-piezo-dielectric matrix[3] is written as

$$\mathrm{P}_{\mathrm{hex}} = \left(\mathrm{P}_{\mathrm{hex}}\right)^T = \left[\begin{array}{cccccc|ccc} A & N & F & 0 & 0 & 0 & 0 & 0 & e_{23} \\ N & A & F & 0 & 0 & 0 & 0 & 0 & e_{23} \\ F & F & C & 0 & 0 & 0 & 0 & 0 & e_{33} \\ 0 & 0 & 0 & L & 0 & 0 & e_{41} & e_{42} & 0 \\ 0 & 0 & 0 & 0 & L & 0 & e_{42} & -e_{41} & 0 \\ 0 & 0 & 0 & 0 & 0 & \frac{A-N}{2} & 0 & 0 & 0 \\ \hline 0 & 0 & 0 & e_{41} & e_{42} & 0 & -\varepsilon_{22} & 0 & 0 \\ 0 & 0 & 0 & e_{42} & -e_{41} & 0 & 0 & -\varepsilon_{22} & 0 \\ e_{23} & e_{23} & e_{33} & 0 & 0 & 0 & 0 & 0 & -\varepsilon_{33} \end{array}\right], \tag{7}$$

[3] A piezoelectric medium of hexagonal symmetry is proved to be transversely isotropic with its sixfold axis being an axis of rotational symmetry.

where the superscript T denotes transposition and we have put

$$C_{11} = A, \quad C_{12} = N, \quad C_{13} = F, \quad C_{33} = C, \quad C_{44} = L.$$

The positivity conditions (2) are equivalent to

$$L > 0, \quad A - N > 0, \quad A + C + N > 0, \quad (A + N)C > 2F^2, \quad \varepsilon_{22} > 0, \quad \varepsilon_{33} > 0.$$

When the sixfold axis lies on the surface of the piezoelectric half-space, a subsonic surface wave called the Bleustein–Gulyaev (BG) wave propagates along that surface in the direction perpendicular to the sixfold axis. Let us take $\mathbf{m} = (1, 0, 0)$ and $\mathbf{n} = (0, 1, 0)$.

Theorem 1 (Bleustein–Gulyaev (BG) wave [2, 5]). *Suppose that $e_{42} \neq 0$. There exists a surface wave which propagates along the surface $x_2 = 0$ of the piezoelectric half-space $x_2 \leq 0$ in the direction of the 1-axis and which satisfies a mechanically free and electrically closed condition at the boundary, i.e., $(\sigma_{i2})_{i\downarrow 1,2,3} = \mathbf{0}$ and $\phi = 0$ at $x_2 = 0$. Its phase velocity $v = v_B^{\text{hex}}$ is given by*

$$V_B^{\text{hex}} = \rho \left(v_B^{\text{hex}}\right)^2 = \frac{L\left(L + 2\dfrac{e_{42}^2}{\varepsilon_{22}}\right)}{L + \dfrac{e_{42}^2}{\varepsilon_{22}}}. \tag{8}$$

The setting of the theorem also implies the existence of another surface wave whose property is determined only from the elastic part $(C_{\alpha\beta})_{\alpha,\beta=1,2,\cdots,6}$ of P_{hex}, i.e., the upper left hand 6×6 block of (7). This is a surface wave whose solution (4) has the fourth component being zero (i.e., $\phi = 0$), and agrees with the Rayleigh wave in a transversely isotropic elastic medium. In fact, Lothe and Barnett [9, 10] proved that

- For fixed orthogonal unit vectors $\mathbf{m}$ and $\mathbf{n}$ in $\mathbb{R}^3$, there are at most two surface waves that satisfy the mechanically free and electrically closed condition at the boundary.

and the argument therein implies that

- Existence condition of the surface waves is stable under a perturbation of material constants.

We are now interested in surface waves which carry information on the piezoelectric tensor $\boldsymbol{e}$, because this tensor makes electric fields and mechanical deformations interact with each other. To the best of the authors' knowledge, however, there are no surface waves, aside from BG waves, whose velocity carries clear information on the piezoelectric tensor and can be used as observation data for recovering the material tensors. On the other hand, the preceding theorem shows that the phase velocity of

BG waves has very restrictive information on the material tensors; it depends on just one component L of $\mathbf{C}$, one component e_{42} of $\boldsymbol{e}$, and one component ε_{22} of $\boldsymbol{\varepsilon}$ when the waves propagate along the surface of a piezoelectric half-space of hexagonal symmetry. Hence in the next section, we consider the perturbation of the phase velocity of BG waves when we add a perturbation to the three tensors $\mathbf{C}$, $\boldsymbol{e}$, and $\boldsymbol{\varepsilon}$ of hexagonal symmetry in P_{hex}.

4 First-Order Perturbation Formula of BG-Wave Velocity

Now we give a perturbation to the material constants of the piezoelectric medium of hexagonal symmetry whose elasto-piezo-dielectric matrix is P_{hex} (7). By using the contracted notation in Sect. 2, this perturbation can be expressed as a perturbative elasto-piezo-dielectric matrix P_{ptb}:

$$\mathrm{P}_{\text{ptb}} = \left(\mathrm{P}_{\text{ptb}}\right)^T = \left[\begin{array}{cccccc|ccc} a_{11} & a_{12} & a_{13} & a_{14} & a_{15} & a_{16} & f_{11} & f_{12} & f_{13} \\ & a_{22} & a_{23} & a_{24} & a_{25} & a_{26} & f_{21} & f_{22} & f_{23} \\ & & a_{33} & a_{34} & a_{35} & a_{36} & f_{31} & f_{32} & f_{33} \\ & & & a_{44} & a_{45} & a_{46} & f_{41} & f_{42} & f_{43} \\ & & & & a_{55} & a_{56} & f_{51} & f_{52} & f_{53} \\ Sym. & & & & & a_{66} & f_{61} & f_{62} & f_{63} \\ \hline & & & & & & -\delta_{11} & -\delta_{12} & -\delta_{13} \\ & & & & & & & -\delta_{22} & -\delta_{23} \\ & & & & & & & & -\delta_{33} \end{array}\right].$$

Namely, the upper left hand 6×6 block is the perturbative part of the elasticity tensor $\mathbf{C}$, the upper right hand 6×3 block is the perturbative part of the piezoelectric tensor $\boldsymbol{e}$, and the lower right hand 3×3 block is the minus[4] of the perturbative part of the dielectric tensor $\boldsymbol{\varepsilon}$, for which we do not assume any material symmetry. Hence the 45 components in the upper triangular part of matrix P_{ptb} are generally all independent.

We consider BG waves in a piezoelectric half-space whose elasto-piezo-dielectric matrix is given by $\mathrm{P}_{\text{hex}} + \mathrm{P}_{\text{ptb}}$, and present a perturbation formula which shows how P_{ptb} affects the phase velocity of BG waves from its comparative value $v_{\mathrm{B}}^{\text{hex}}$ for the piezoelectric half-space of hexagonal symmetry.

Theorem 2 (Perturbation of BG waves). *Suppose that* $v_{\mathrm{B}}^{\text{hex}} \neq v_{\mathrm{R}}$, *where* v_{R} *is the velocity of Rayleigh waves determined from the upper left hand* 6×6 *block in* P_{hex}. *In*

[4]The minus signs of δ_{ij} in the lower right hand 3×3 block come from the minus sign on the right-hand side of the constitutive equation for the electric displacement $\boldsymbol{D}$ in (1). See the lower right hand 3×3 block of the 9×9 elasto-piezo-dielectric matrix appearing at the last paragraph of Sect. 2 and see also the same block of the elasto-piezo-dielectric matrix for the piezoelectric medium of hexagonal symmetry (7). Note that the (i, j) components of the dielectric tensor pertaining to the perturbed piezoelectric medium are written as $\varepsilon_{ij} + \delta_{ij}$ with $\varepsilon_{11} = \varepsilon_{22}$ and $\varepsilon_{ij} = 0$ $(i \neq j)$.

a piezoelectric medium whose elasto-piezo-dielectric matrix is given by $\mathrm{P}_{\mathrm{hex}} + \mathrm{P}_{\mathrm{ptb}}$, *the phase velocity* v_{B} *of BG waves which propagate along the surface* $x_2 = 0$ *of the piezoelectric half-space* $x_2 \leq 0$ *in the direction of the* 1*-axis and which satisfies a mechanically free and electrically closed condition at* $x_2 = 0$ *can be written, to within terms linear in the perturbative part* $\mathrm{P}_{\mathrm{ptb}}$, *as*

$$V_{\mathrm{B}} = \rho\,(v_{\mathrm{B}})^2 = V_{\mathrm{B}}^{\mathrm{hex}} + P_1\, f_{42} + P_2\, f_{51} + D_1\, \delta_{11} + D_2\, \delta_{22} + E_1\, a_{44} + E_2\, a_{55},$$

where $V_{\mathrm{B}}^{\mathrm{hex}}$ *is given by (8) and the coefficients* P_i, D_i *and* E_i $(i = 1, 2)$ *are written by*

$$P_1 = \frac{-2\,L\, {e_{42}}^5}{(L\,\varepsilon_{22} + {e_{42}}^2)^2\,(L\,\varepsilon_{22} + 2\,{e_{42}}^2)}, \qquad P_2 = \frac{2\,L\, e_{42}}{L\,\varepsilon_{22} + 2\,{e_{42}}^2},$$

$$D_1 = \frac{-L^2\, {e_{42}}^2}{(L\,\varepsilon_{22} + {e_{42}}^2)\,(L\,\varepsilon_{22} + 2\,{e_{42}}^2)}, \qquad D_2 = \frac{-L^2\, {e_{42}}^4}{(L\,\varepsilon_{22} + {e_{42}}^2)^2\,(L\,\varepsilon_{22} + 2\,{e_{42}}^2)},$$

$$E_1 = \frac{{e_{42}}^4}{(L\,\varepsilon_{22} + {e_{42}}^2)^2}, \qquad E_2 = 1.$$

Remark 1 Only two components f_{42}, f_{51} of the perturbative part of the piezoelectric tensor $\boldsymbol{e}$, two components δ_{11}, δ_{22} of the perturbative part of the dielectric tensor $\boldsymbol{\varepsilon}$ and two components a_{44}, a_{55} of the perturbative part of the elasticity tensor $\mathbf{C}$ can affect the first-order perturbation of the phase velocity v_{B} of BG waves that propagate in the direction of the 1-axis on the surface of the half-space $x_2 \leq 0$.

Remark 2 The components of $\mathrm{P}_{\mathrm{hex}}$ remain invariant under the rotation of the medium around the 3-axis (cf. footnote 3). Hence the transformation formulas for the tensors $\mathbf{C}$, $\boldsymbol{e}$ and $\boldsymbol{\varepsilon}$ can be applied to observe that the phase velocity v_{B}^{θ} of BG waves which propagate along the surface $-x_1 \sin\theta + x_2 \cos\theta = 0$ of the piezoelectric half-space $-x_1 \sin\theta + x_2 \cos\theta \leq 0$ in the direction of $(\cos\theta, \sin\theta, 0)$ can be written, to within terms linear in the perturbative part $\mathrm{P}_{\mathrm{ptb}}$, as

$$\begin{aligned} V_{\mathrm{B}}^{\theta} = \rho\,(v_{\mathrm{B}}^{\theta})^2 = V_{\mathrm{B}}^{\mathrm{hex}} &+ \cos^2\theta\Big(P_1\, f_{42} + P_2\, f_{51} + D_1\, \delta_{11} + D_2\, \delta_{22} + E_1\, a_{44} + E_2\, a_{55}\Big) \\ &+ \cos\theta \sin\theta\,\Big((P_2 - P_1)(f_{52} + f_{41}) + 2(D_1 - D_2)\delta_{12} + 2(E_2 - E_1)a_{45}\Big) \\ &+ \sin^2\theta\,\Big(P_2\, f_{42} + P_1\, f_{51} + D_2\, \delta_{11} + D_1\, \delta_{22} + E_2\, a_{44} + E_1\, a_{55}\Big), \end{aligned}$$

where $V_{\mathrm{B}}^{\mathrm{hex}}$, the coefficients P_i, D_i and E_i $(i = 1, 2)$ are the same as depicted in the theorem.

In the next section, we briefly summarize the Barnett–Lothe integral formalism for piezoelectricity [9, 10], which is the essence of the proof of Theorem 2.

5 Integral Formalism and Secular Equation

For the orthogonal unit vectors $\mathbf{m} = (m_1, m_2, m_3)$ and $\mathbf{n} = (n_1, n_2, n_3)$ in $\mathbb{R}^3$ (see the paragraph in Sect. 2 which includes formula (4)), define the 4×4 real matrices $\mathbf{Q}$, $\mathbf{R}$ and $\mathbf{T}$ blockwise as follows:

$$\mathbf{Q} = \left(\begin{array}{c|c} \left(\sum_{j,l=1}^{3} C_{ijkl} m_j m_l\right)_{i\downarrow k\to 1,2,3} - \rho v^2 \mathbf{I} & \left(\sum_{j,l=1}^{3} e_{ijl} m_j m_l\right)_{i\downarrow 1,2,3} \\ \hline \left(\sum_{j,l=1}^{3} e_{klj} m_j m_l\right)_{k\to 1,2,3} & -\sum_{j,l=1}^{3} \varepsilon_{jl} m_j m_l \end{array} \right),$$

$$\mathbf{R} = \left(\begin{array}{c|c} \left(\sum_{j,l=1}^{3} C_{ijkl} m_j n_l\right)_{i\downarrow k\to 1,2,3} & \left(\sum_{j,l=1}^{3} e_{ijl} m_j n_l\right)_{i\downarrow 1,2,3} \\ \hline \left(\sum_{j,l=1}^{3} e_{klj} m_j n_l\right)_{k\to 1,2,3} & -\sum_{j,l=1}^{3} \varepsilon_{jl} m_j n_l \end{array} \right),$$

$$\mathbf{T} = \left(\begin{array}{c|c} \left(\sum_{j,l=1}^{3} C_{ijkl} n_j n_l\right)_{i\downarrow k\to 1,2,3} & \left(\sum_{j,l=1}^{3} e_{ijl} n_j n_l\right)_{i\downarrow 1,2,3} \\ \hline \left(\sum_{j,l=1}^{3} e_{klj} n_j n_l\right)_{k\to 1,2,3} & -\sum_{j,l=1}^{3} \varepsilon_{jl} n_j n_l \end{array} \right), \tag{9}$$

where $(M_{ik})_{i\downarrow k\to 1,2,3}$ denotes a 3×3 matrix whose (i, k) component is M_{ik}, $(v_i)_{i\downarrow 1,2,3}$ a column vector in $\mathbb{R}^3$ whose i-th component is v_i, $(w_k)_{k\to 1,2,3}$ a row vector in $\mathbb{R}^3$ whose k-th component is w_k, and $\mathbf{I}$ the 3×3 identity matrix.

Substituting (4) into (1), we can compute the vectors $\mathbf{l}_\alpha \in \mathbb{C}^4$ $(1 \le \alpha \le 4)$ so that

$$\left(\begin{array}{c} \left(\sum_{j=1}^{3} \sigma_{ij} n_j\right)_{i\downarrow 1,2,3} \\ \sum_{i=1}^{3} D_i n_i \end{array} \right)_{\mathbf{n}\cdot\mathbf{x}=0} = -\mathrm{i}\, k \sum_{\alpha=1}^{4} c_\alpha\, \mathbf{l}_\alpha\, \mathrm{e}^{-\mathrm{i}\, k(\mathbf{m}\cdot\mathbf{x} - v\, t)}. \tag{10}$$

Then the column vectors $\begin{pmatrix} \mathbf{a}_\alpha \\ \mathbf{l}_\alpha \end{pmatrix} \in \mathbb{C}^8$ $(1 \le \alpha \le 4)$, whose first four components consist of $\mathbf{a}_\alpha$ in (4) and whose last three components consist of $\mathbf{l}_\alpha$ in (10), and $p_\alpha \in \mathbb{C}$ $(\mathrm{Im}\, p_\alpha > 0, 1 \le \alpha \le 4)$ become eigenvectors and eigenvalues of Stroh's eight-dimensional eigenvalue problem, respectively [9, 10] :

$$\mathbf{N} \begin{pmatrix} \mathbf{a}_\alpha \\ \mathbf{l}_\alpha \end{pmatrix} = p_\alpha \begin{pmatrix} \mathbf{a}_\alpha \\ \mathbf{l}_\alpha \end{pmatrix}, \qquad \alpha = 1, 2, 3, 4, \tag{11}$$

where $\mathbf{N}$ is a 8×8 matrix defined by

$$\mathbf{N} = \begin{pmatrix} -\mathbf{T}^{-1}\mathbf{R}^T & \mathbf{T}^{-1} \\ -\mathbf{Q} + \mathbf{R}\mathbf{T}^{-1}\mathbf{R}^T & -\mathbf{R}\mathbf{T}^{-1} \end{pmatrix}. \tag{12}$$

Now let $\widetilde{\mathbf{m}} = (\widetilde{m}_1, \widetilde{m}_2, \widetilde{m}_3)$ and $\widetilde{\mathbf{n}} = (\widetilde{n}_1, \widetilde{n}_2, \widetilde{n}_3)$ be orthogonal unit vectors in $\mathbb{R}^3$ which are obtained by rotating the orthogonal unit vectors $\mathbf{m}$ and $\mathbf{n}$ around their vector product $\mathbf{m} \times \mathbf{n}$ by an angle ϕ $(-\pi \le \phi < \pi)$ so that

$$\widetilde{\mathbf{m}} = \widetilde{\mathbf{m}}(\phi) = \mathbf{m}\cos\phi + \mathbf{n}\sin\phi, \qquad \widetilde{\mathbf{n}} = \widetilde{\mathbf{n}}(\phi) = -\mathbf{m}\sin\phi + \mathbf{n}\cos\phi.$$

Let $\mathbf{Q}(\phi)$, $\mathbf{R}(\phi)$ and $\mathbf{T}(\phi)$ be the 4×4 real matrices given blockwise as follows:

$$\mathbf{Q}(\phi) = \left(\begin{array}{c|c} \left(\sum_{j,l=1}^{3} C_{ijkl}\widetilde{m}_j\widetilde{m}_l\right)_{i\downarrow k\to 1,2,3} - \rho v^2\cos^2\phi\,\mathbf{I} & \left(\sum_{j,l=1}^{3} e_{ijl}\widetilde{m}_j\widetilde{m}_l\right)_{i\downarrow 1,2,3} \\ \hline \left(\sum_{j,l=1}^{3} e_{klj}\widetilde{m}_j\widetilde{m}_l\right)_{k\to 1,2,3} & -\sum_{j,l=1}^{3} \varepsilon_{jl}\widetilde{m}_j\widetilde{m}_l \end{array}\right),$$

$$\mathbf{R}(\phi) = \left(\begin{array}{c|c} \left(\sum_{j,l=1}^{3} C_{ijkl}\widetilde{m}_j\widetilde{n}_l\right)_{i\downarrow k\to 1,2,3} + \rho v^2\cos\phi\sin\phi\,\mathbf{I} & \left(\sum_{j,l=1}^{3} e_{ijl}\widetilde{m}_j\widetilde{n}_l\right)_{i\downarrow 1,2,3} \\ \hline \left(\sum_{j,l=1}^{3} e_{klj}\widetilde{m}_j\widetilde{n}_l\right)_{k\to 1,2,3} & -\sum_{j,l=1}^{3} \varepsilon_{jl}\widetilde{m}_j\widetilde{n}_l \end{array}\right),$$

$$\mathbf{T}(\phi) = \left(\begin{array}{c|c} \left(\sum_{j,l=1}^{3} C_{ijkl}\widetilde{n}_j\widetilde{n}_l\right)_{i\downarrow k\to 1,2,3} - \rho v^2\sin^2\phi\,\mathbf{I} & \left(\sum_{j,l=1}^{3} e_{ijl}\widetilde{n}_j\widetilde{n}_l\right)_{i\downarrow 1,2,3} \\ \hline \left(\sum_{j,l=1}^{3} e_{klj}\widetilde{n}_j\widetilde{n}_l\right)_{k\to 1,2,3} & -\sum_{j,l=1}^{3} \varepsilon_{jl}\widetilde{n}_j\widetilde{n}_l \end{array}\right). \tag{13}$$

Note that $\mathbf{Q}(0)$, $\mathbf{R}(0)$ and $\mathbf{T}(0)$ are equal to $\mathbf{Q}$, $\mathbf{R}$ and $\mathbf{T}$ in (9), respectively.

The limiting velocity $v_L = v_L(\mathbf{m}, \mathbf{n})$ is the lowest velocity for which the matrix $\mathbf{T}(\phi)$ becomes singular for some angle ϕ:

$$v_L = \inf\{v > 0 \mid \exists\phi;\ \det\mathbf{T}(\phi) = 0\}.$$

The interval $0 < v < v_L$ is called the subsonic range. Surface waves whose phase velocity is included in the subsonic range are called subsonic surface waves.

Like (12), we define a 8×8 matrix $\mathbf{N}(\phi)$ by

$$\mathbf{N}(\phi) = \begin{pmatrix} -\mathbf{T}(\phi)^{-1}\mathbf{R}(\phi)^T & \mathbf{T}(\phi)^{-1} \\ -\mathbf{Q}(\phi) + \mathbf{R}(\phi)\mathbf{T}(\phi)^{-1}\mathbf{R}(\phi)^T & -\mathbf{R}(\phi)\mathbf{T}(\phi)^{-1} \end{pmatrix} \tag{14}$$

for $0 \le v < v_L$. Then $\mathbf{N}(0)$ is equal to $\mathbf{N}$ of (12).

One of the essentials of the Stroh formalism [1, 3, 8, 14, 18] is that the eigenvectors $\begin{pmatrix} \mathbf{a}_\alpha \\ \mathbf{l}_\alpha \end{pmatrix} \in \mathbb{C}^8$ $(1 \le \alpha \le 4)$ of $\mathbf{N} = \mathbf{N}(0)$ remain to be eigenvectors of $\mathbf{N}(\phi)$ for all ϕ, i.e.,

$$\mathbf{N}(\phi)\begin{pmatrix} \mathbf{a}_\alpha \\ \mathbf{l}_\alpha \end{pmatrix} = p_\alpha(\phi)\begin{pmatrix} \mathbf{a}_\alpha \\ \mathbf{l}_\alpha \end{pmatrix}, \qquad \alpha = 1, 2, 3, 4 \tag{15}$$

for all ϕ, whereas $p_\alpha(\phi)$ is the solution to $p_\alpha'(\phi) = -1 - p_\alpha(\phi)^2$ with $p_\alpha(0) = p_\alpha$.

We take the angular average of both sides of the eigenrelation (15) with respect to ϕ over $[-\pi, \pi]$, allowing for $\frac{1}{2\pi}\int_{-\pi}^{\pi} p_\alpha(\phi)\, d\phi = \mathrm{i}$ (see the references listed in the preceding paragraph), to obtain

$$\begin{pmatrix} \mathbf{S}_1 & \mathbf{S}_2 \\ \mathbf{S}_3 & \mathbf{S}_1^T \end{pmatrix} \begin{pmatrix} \mathbf{a}_\alpha \\ \mathbf{l}_\alpha \end{pmatrix} = \mathrm{i} \begin{pmatrix} \mathbf{a}_\alpha \\ \mathbf{l}_\alpha \end{pmatrix}, \qquad \alpha = 1, 2, 3, 4, \tag{16}$$

where $\mathbf{S}_i = \mathbf{S}_i(v)$ $(i = 1, 2, 3)$ are the 4×4 real matrices, each of which is the angular average of the 4×4 blocks in the matrix $\mathbf{N}(\phi)$[5]:

$$\mathbf{S}_1 = \frac{1}{2\pi}\int_{-\pi}^{\pi} -\mathbf{T}(\phi)^{-1}\mathbf{R}(\phi)^T\, d\phi, \qquad \mathbf{S}_2 = \frac{1}{2\pi}\int_{-\pi}^{\pi} \mathbf{T}(\phi)^{-1}\, d\phi,$$
$$\mathbf{S}_3 = \frac{1}{2\pi}\int_{-\pi}^{\pi} -\mathbf{Q}(\phi) + \mathbf{R}(\phi)\mathbf{T}(\phi)^{-1}\mathbf{R}(\phi)^T\, d\phi. \tag{17}$$

Let us turn to the mechanically free and electrically closed boundary condition (6). This condition implies through (4) and (10) that there exists a nontrivial set of $c_\alpha \in \mathbb{C}$ $(1 \leq \alpha \leq 4)$ so that

$$\sum_{\alpha=1}^{4} c_\alpha \begin{pmatrix} (\mathbf{a}_\alpha)_4 \\ (\mathbf{l}_\alpha)_1 \\ (\mathbf{l}_\alpha)_2 \\ (\mathbf{l}_\alpha)_3 \end{pmatrix} = \mathbf{0} \quad \text{at } v = v_{\mathrm{B}}, \tag{18}$$

where $(\mathbf{v})_i$ denotes the i-th component of a vector $\mathbf{v} \in \mathbb{C}^4$. Taking a linear combination of both sides of (16) with the coefficients $c_\alpha \in \mathbb{C}$ $(1 \leq \alpha \leq 4)$ in (18), we have

$$\begin{pmatrix} \mathbf{S}_1 & \mathbf{S}_2 \\ \mathbf{S}_3 & \mathbf{S}_1^T \end{pmatrix} \sum_{\alpha=1}^{4} c_\alpha \begin{pmatrix} \mathbf{a}_\alpha \\ \mathbf{l}_\alpha \end{pmatrix} = \mathrm{i} \sum_{\alpha=1}^{4} c_\alpha \begin{pmatrix} \mathbf{a}_\alpha \\ \mathbf{l}_\alpha \end{pmatrix}.$$

We extract the 4th to 7th components from both sides of the preceding system and use (18) to obtain a four-dimensional linear system

$$\begin{pmatrix} (\mathbf{S}_1)_{\substack{i=4 \\ k\to 1,2,3}} & (\mathbf{S}_2)_{\substack{i=4 \\ k=4}} \\ (\mathbf{S}_3)_{\substack{i\downarrow 1,2,3 \\ k\to 1,2,3}} & (\mathbf{S}_1^T)_{\substack{i\downarrow 1,2,3 \\ k=4}} \end{pmatrix} \sum_{\alpha=1}^{4} c_\alpha \begin{pmatrix} (\mathbf{a}_\alpha)_1 \\ (\mathbf{a}_\alpha)_2 \\ (\mathbf{a}_\alpha)_3 \\ (\mathbf{l}_\alpha)_4 \end{pmatrix} = \mathbf{0} \quad \text{at } v = v_{\mathrm{B}}, \tag{19}$$

[5]When the eigenvalue problem (11) is degenerate and generalized eigenvectors must be introduced, the general form of the solutions (4) will have to be slightly modified. In that case, however, the relation (16) still holds not only for eigenvectors but also for generalized eigenvectors of (11) corresponding to the eigenvalues p_α $(1 \leq \alpha \leq 4)$ of positive imaginary part. For details, we again refer to the literature listed in the preceding paragraph.

where each block component of the 4×4 matrix on the left-hand side denotes a submatrix which consists of the (i, k) component of the parenthesized matrix with i and k moving through the range described therein. For the system (19) to have a nontrivial solution we claim that

$$\det \begin{pmatrix} (\mathbf{S}_1)_{\substack{i=4 \\ k\to 1,2,3}} & (\mathbf{S}_2)_{\substack{i=4 \\ k=4}} \\ (\mathbf{S}_3)_{\substack{i\downarrow 1,2,3 \\ k\to 1,2,3}} & (\mathbf{S}_1^T)_{\substack{i\downarrow 1,2,3 \\ k=4}} \end{pmatrix} = 0 \quad \text{at } v = v_{\mathrm{B}}. \tag{20}$$

This is a secular equation for the phase velocity of BG waves [9, 10]. We observe from (13) and (17) that the 4×4 matrix on the left-hand side is a function of the material constants and velocity v. Hence we can apply a perturbation method to (20) to obtain Theorem 2. The details will be given in our forthcoming paper [12].

6 Concluding Remarks

In this paper, we give rudimentary accounts of surface waves in a piezoelectric half-space, putting emphasis on the Bleustein–Gulyaev (BG) waves that propagate along the surface of a piezoelectric half-space of hexagonal symmetry, on which the mechanically free and electrically closed boundary condition is imposed. We add a fully anisotropic perturbation to the three material tensors of the hexagonal piezoelectric half-space, i.e., to the elasticity tensor, the piezoelectric tensor and to the dielectric tensor, each of which has a hexagonal symmetry. We then study the direct problem of deriving a first-order perturbation formula for the phase velocity of BG waves, which expresses the shift in the velocity from its comparative value for a hexagonal piezoelectric half-space, caused by the perturbative parts of the aforementioned three material tensors. It can be observed that only a few components of the perturbative parts of those tensors can affect the first-order perturbation of the phase velocity of BG waves. We give an outline of the proof of the perturbation formula, where the Barnett–Lothe integral formalism is essential.

Acknowledgments We thank an anonymous reviewer for his/her careful reading of our manuscript and for the helpful comments. The work of Tanuma was partly supported by JSPS KAKENHI Grant Numbers JP26400157, JP19K03559. The research efforts of Xu were partially supported by NSFC 11471284, 11621101, 91630309.

References

1. Barnett, D.M., Lothe, J.: Synthesis of the sextic and the integral formalism for dislocations, Green's functions, and surface waves in anisotropic elastic solids. Phys. Norv. **7**, 13–19 (1973)
2. Bleustein, J.L.: A new surface wave in piezoelectric materials. Appl. Phys. Lett. **13**, 412–413 (1968)

3. Chadwick, P., Smith, G.D.: Foundations of the theory of surface waves in anisotropic elastic materials. Adv. Appl. Mech. **17**, 303–376 (1977)
4. Gualtieri, J.G., Kosinski, J.A., Ballato, A.: Piezoelectric materials for acoustic wave applications. IEEE Trans. Ultrason., Ferroelec., Freq. Contr. **41**, 53–59 (1994)
5. Gulyaev, YuV: Electroacoustic surface waves in solids. Sov. Phys. JETP Lett. **9**, 37–38 (1969)
6. Kaplunov, J., Prikazchikov, D.A.: Asymptotic theory for Rayleigh and Rayleigh-type waves. Adv. Appl. Mech. **50**, 1–109 (2017)
7. Kaplunov, J., Zakharov, A., Prikazchikov, D.: Explicit models for elastic and piezoelastic surface waves. IMA J. Appl. Math. **71**, 768–782 (2006)
8. Lothe, J., Barnett, D.M.: On the existence of surface-wave solutions for anisotropic elastic half-spaces with free surface. J. Appl. Phys. **47**, 428–433 (1976)
9. Lothe, J., Barnett, D.M.: Integral formalism for surface waves in piezoelectric crystals. Existence considerations. J. Appl. Phys. **47**, 1799–1807 (1976)
10. Lothe, J., Barnett, D.M.: Further development of the theory for surface waves in piezoelectric crystals. Phys. Norv. **8**, 239–254 (1976)
11. Maugin, G.A.: Continuum Mechanics of Electromagnetic Solids. North-Holland, Amsterdam (1988)
12. Nakamura, G., Tanuma, K., Xu, X.: Perturbation formula for phase velocity of Bleustein-Gulyaev waves in piezoelectric media (in preparation)
13. Ristic, V.M.: Principles of Acoustic Devices. Wiley, New York (1983)
14. Tanuma, K.: Stroh formalism and Rayleigh waves. J. Elasticity **89**, 5–154 (2007)
15. Tanuma, K., Man, C.-S.: Perturbation formula for phase velocity of Rayleigh waves in prestressed anisotropic media. J. Elasticity **85**, 21–37 (2006)
16. Tanuma, K., Man, C.-S., Du, W.: Perturbation of phase velocity of Rayleigh waves in prestressed anisotropic media with orthorhombic principal part. Math. Mech. Solids **18**, 301–322 (2013)
17. Tiersten, H.F.: Linear Piezoelectric Plate Vibrations. Plenum Press, New York (1969)
18. Ting, T.C.T.: Anisotropic Elasticity. Oxford University Press, New York (1996)

Deformation Problem for Glued Elastic Bodies and an Alternative Iteration Method

Masato Kimura and Atsushi Suzuki

Abstract We study a mathematical model for deformation of glued elastic bodies in 2D or 3D, which is a linear elasticity system with adhesive force on the glued surface. We reveal a variational structure of the model and prove the unique existence of a weak solution based on it. Furthermore, we also consider an alternating iteration method and show that it is nothing but an alternating minimizing method of the total energy. The convergence to a monolithic formulation and the alternating iteration method are numerically studied with the finite element method.

1 Introduction

We consider a mathematical model which describes deformation of two elastic bodies glued to each other on a surface. The understanding of such glued structure or adhesive bonding process is important in industrial and scientific applications, especially in the case that the mechanical bonding technique exhibits its disadvantages comparing with the adhesive one, e.g., bonding between soft materials, or one between very small-scaled materials. The importance of mathematical modeling and numerical simulation is increasing in the design of desirable mechanical properties of composite materials with glued layer structure.

In mathematics, mathematical models of Frémond [3] and Roubíček et al. [4] proposed mathematical models of such glued structure and its delamination process is redundant. Scala [5] studied more extended delamination model with kinetic and viscoelastic terms and proved existence of a solution. For further mathematical studies on the delamination process, we refer the above works and references therein.

M. Kimura (✉)
Faculty of Mathematics and Physics, Kanazawa University, Kakuma, Kanazawa 920-1192, Japan
e-mail: mkimura@se.kanazawa-u.ac.jp

A. Suzuki
Cybermedia Center, Osaka University, Machikaneyama, Toyonaka, Osaka 560-0043, Japan
e-mail: atsushi.suzuki@cas.cmc.osaka-u.ac.jp

H. Itou et al. (eds.), *Mathematical Analysis of Continuum Mechanics and Industrial Applications III*, Mathematics for Industry 34,
https://doi.org/10.1007/978-981-15-6062-0_6

In this paper, we concentrate on the stationary deformation problem of the glued structure, which is a linear elasticity system with adhesive force on the glued surface. A similar stationary problem also appears in the implicit time discretization of the above delamination models [7].

The outline of this paper is as follows. In Sect. 2, we describe the deformation model and give a definition of a weak solution. Section 3 is devoted to review several known consequences from the coercivity of a bilinear form; the existence and the uniqueness of a weak solution, a variational principle, and an error estimate of a finite element approximation.

For the purpose of efficient numerical computation of the obtained weak form of our model in 2D and 3D, we propose an alternating iteration method in Sect. 4. The alternating iteration method was proposed in [7] and was used to simulate the vibration-delamination model proposed in [5]. We will show that it is nothing but an alternating energy minimization procedure. In particular, it generates a sequence of displacements which monotonically decreases the total energy (Theorem 4.1).

In Sect. 5, we consider finite element discretization. We give discrete forms of the monolithic method and the alternating iteration method and prove that the sequence generated by the alternating iteration method converges to the discrete solution by the monolithic method. Those theoretical results are also supported by numerical experiments in three-dimensional setting.

2 Deformation of Glued Elastic Bodies

We consider a bounded Lipschitz domain $\Omega \subset \mathbb{R}^d$, $(d = 2, 3)$. We suppose that

$$\Omega \setminus \Gamma = \Omega_1 \cup \Omega_2, \quad \Omega_1 \cap \Omega_2 = \emptyset,$$

where Γ is a Lipschitz surface which is the common boundary of two disjoint Lipschitz domains Ω_1 and Ω_2 as shown in Figs. 1 and 2. We denote by ν the unit normal vector on Γ pointing from Ω_1 into Ω_2, and the one on $\partial\Omega$ pointing outward. We suppose that the boundary $\partial\Omega$ is decomposed to the following disjoint portions:

$$\partial\Omega = \partial_D\Omega \cup \overline{\partial_N\Omega}, \quad \partial_D\Omega \cap \overline{\partial_N\Omega} = \emptyset, \quad \mathcal{H}^{d-1}(\overline{\partial_N\Omega} \setminus \partial_N\Omega) = 0,$$

where $\partial_D\Omega$ and $\partial_N\Omega$ are relatively open subsets of $\partial\Omega$ and $\mathcal{H}^{d-1}$ denotes the $d-1$ dimensional Hausdorff measure. We also define $\partial_D\Omega_i := \partial_D\Omega \cap \partial\Omega_i$ and $\partial_N\Omega_i := \partial_N\Omega \cap \partial\Omega_i$ for $i = 1, 2$. We assume that $\partial_D\Omega$ is a nonempty relatively open subset of $\partial\Omega$, and $\mathcal{H}^{d-1}(\partial_D\Omega_1)$ or $\mathcal{H}^{d-1}(\partial_D\Omega_2)$ is positive. Without loss of generality, we always suppose $\mathcal{H}^{d-1}(\partial_D\Omega_1) > 0$ (Figs. 1 and 2).

In this paper, we consider the following stationary deformation model of two elastic bodies Ω_1 and Ω_2 that are glued by an adhesive on Γ. We consider an adhesion force but ignore a friction force on the interface. The problem is to find $u : \Omega \setminus \Gamma \to \mathbb{R}^d$ such that:

Fig. 1 $\partial_D\Omega \subset \partial\Omega_1 \cup \partial\Omega_2$

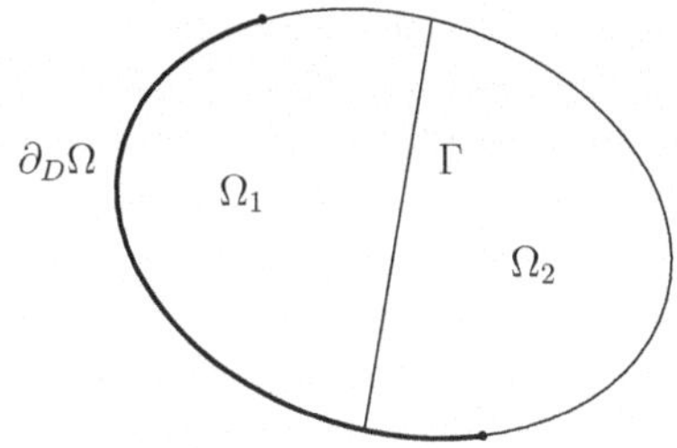

Fig. 2 $\partial_D\Omega \subset \partial\Omega_1$

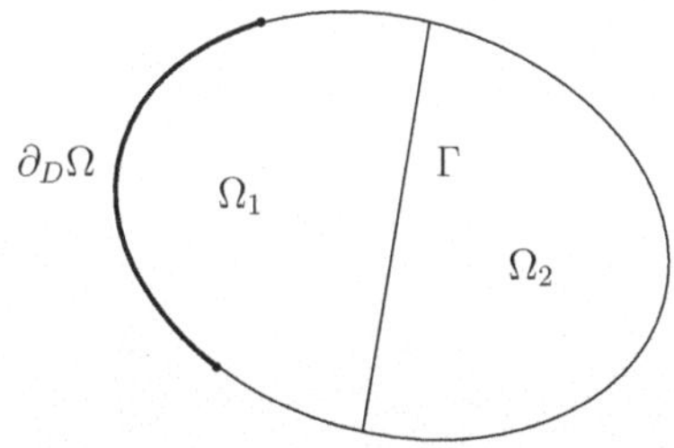

$$\begin{cases} -\operatorname{div}\sigma(u) = f(x) & (x \in \Omega \setminus \Gamma), \\ u = g(x) & (x \in \partial_D\Omega), \\ \sigma(u)\nu = q(x) & (x \in \partial_N\Omega), \\ \sigma(u_1)\nu = \zeta(x)[u] = \sigma(u_2)\nu & (x \in \Gamma). \end{cases} \tag{1}$$

The meanings of the above symbols are as follows. We use the Einstein summation convention in this section. For matrices $\xi = (\xi_{kl}),\ \eta = (\eta_{kl}) \in \mathbb{R}^{d\times d}$, we denote their component-wise inner product by $\xi : \eta := \xi_{kl}\eta_{kl}$ and the norm by $|\xi| := \sqrt{\xi : \xi}$.

The solution u is a displacement field on $\Omega \setminus \Gamma = \Omega_1 \cup \Omega_2$. We denote $u|_{\Omega_i}$ by u_i for $i = 1, 2$, and often write $u = (u_1, u_2)$. The symmetric gradient of u is defined by

$$e(u) := \frac{1}{2}\left(\nabla u^{\mathrm{T}} + \left(\nabla u^{\mathrm{T}}\right)^{\mathrm{T}}\right) \in \mathbb{R}^{d\times d}_{\mathrm{sym}}.$$

The stress tensor $\sigma(u) \in \mathbb{R}^{d\times d}_{\mathrm{sym}}$ satisfies the constitutive relation

$$\sigma(u) := Ce(u) = (c_{klmn}\, e_{mn}(u))_{kl} \in \mathbb{R}^{d\times d}_{\mathrm{sym}},$$

where $C := C(x) = (c_{klmn}(x)) \in \mathbb{R}^{d\times d\times d\times d}$ is the elasticity tensor with the symmetries $c_{klmn} = c_{mnkl} = c_{lkmn} (1 \le k, l, m, n \le d)$.

We assume that $C \in L^\infty(\Omega; \mathbb{R}^{d\times d\times d\times d})$ and there exists $c_* > 0$ such that

$$c_{klmn}(x)\xi_{kl}\xi_{mn} \ge c_*|\xi|^2 \quad (\text{a.e. } x \in \Omega,\ \xi \in \mathbb{R}^{d\times d}_{\mathrm{sym}}).$$

The first equation of (1) is the force-balance equation in each subdomain Ω_i, where f is a given body force. The second and third equations are Dirichlet and

Neumann boundary conditions, where g is a given displacement on $\partial_D \Omega$ and q is a given surface traction on $\partial_N \Omega$.

In the fourth equation, $\zeta(x) \geq 0$ is a given adhesive parameter on the glued surface Γ which represents the strength of adhesive bonding. The adhesive force at $x \in \Gamma$ is assumed to be $\zeta(x)[u]$, where $[u(x)] := (u_2(x) - u_1(x))$ is the gap of the displacement $u = (u_1, u_2)$ on Γ. It should be balanced with the surface traction force $\sigma(u_1(x))\nu$ on Ω_1 side and also with $\sigma(u_2(x))\nu$ on Ω_2 side.

To consider a weak formulation of (1), we introduce the following spaces,

$$X_i := H^1(\Omega_i; \mathbb{R}^d), \quad V_i := \{u_i \in X_i;\ u_i|_{\partial_D \Omega_i} = 0\} \quad (i = 1, 2),$$
$$X := H^1(\Omega \setminus \Gamma; \mathbb{R}^d) \cong X_1 \times X_2, \quad V := \{u \in X;\ u|_{\partial_D \Omega} = 0\} \cong V_1 \times V_2.$$

For $g = (g_1, g_2) \in X$, we also define affine spaces:

$$V(g) := V + g = \{u \in X;\ u - g \in V\}, \quad V_i(g_i) := V_i + g_i \quad (i = 1, 2).$$

Definition 2.1 (*Weak solution*) We suppose that a Dirichlet boundary data $g \in X$, a body force $f \in L^2(\Omega; \mathbb{R}^d)$, a surface traction $q \in L^2(\partial_N \Omega; \mathbb{R}^d)$, and an adhesive coefficient $\zeta \in L^\infty(\Gamma)$, $\zeta(x) \geq 0$ are given. Then, we call u a weak solution of (1) if

$$u \in V(g), \quad a_0(u, v) = l_0(v) \ \text{ for all } v \in V,$$

where

$$a_0(u, v) := \int_{\Omega \setminus \Gamma} \sigma(u) : e(v)\, dx + \int_\Gamma \zeta[u] \cdot [v]\, ds \quad (u, v \in X), \tag{2}$$
$$l_0(v) := \int_{\Omega \setminus \Gamma} f \cdot v\, dx + \int_{\partial_N \Omega} q \cdot v\, ds \quad (v \in X).$$

The bilinear form a_0 and the linear form l_0 are decomposed into sum of the following subforms:

$$a_0(u, v) = \sum_{i=1}^{2} a_i(u_i, v_i) + a_\Gamma([u], [v]) \quad (u,\ v \in X),$$
$$a_i(u_i, v_i) := \int_{\Omega_i} \sigma(u_i) : e(v_i)\, \mathrm{d}x, \ (i = 1, 2), \quad a_\Gamma(u, v) := \int_\Gamma \zeta u \cdot v\, \mathrm{d}s,$$
$$l_0(v) = \sum_{i=1}^{2} l_i(v_i), \quad l_i(v_i) := \int_{\Omega_i} f \cdot v_i\, \mathrm{d}x + \int_{\partial_N \Omega_i} q \cdot v_i\, \mathrm{d}s \ \ (v_i \in X_i,\ i = 1, 2).$$

We remark that if $c_{klmn} \in C^1(\overline{\Omega_i})$ for $i = 1, 2$, then a strong solution of (1), i.e., $u \in H^2(\Omega \setminus \Gamma; \mathbb{R}^d)$, which satisfies (1) almost everywhere on $\Omega \setminus \Gamma$ or on $\partial\Omega \cup \Gamma$, is a weak solution. On the other hand, if a weak solution belongs to $H^2(\Omega \setminus \Gamma; \mathbb{R}^d)$, then it is a strong solution.

3 Unique Existence of a Weak Solution

For establishing the unique existence of a weak solution, the coercivity of the bilinear form $a_0(u, v)$ defined in (2) is essential.

Theorem 3.1 (Coercivity of a_0) *We suppose that* $\zeta(x) \geq 0$ *and* $\|\zeta\|_{L^\infty(\Gamma)} > 0$. *Then there exists* $a_* > 0$ *such that* $a_0(v, v) \geq a_* \|v\|_X^2$ *holds for all* $v \in V$.

A slightly long proof of this theorem using an argument by contradiction was given in [7] and another simpler proof will be given in our forthcoming paper. We postpone the proof to it and here we just remark that the case of $\mathcal{H}^{d-1}(\partial_D \Omega_i) > 0$ for both $i = 1, 2$ is relatively easily shown by using Körn's second inequality [2].

As a consequence of Theorem 3.1, we immediately have the unique existence of a weak solution.

Theorem 3.2 (Unique existence) *We suppose that Dirichlet boundary data* $g \in X$, *a body force* $f \in L^2(\Omega; \mathbb{R}^d)$, *a surface traction* $q \in L^2(\partial_N \Omega; \mathbb{R}^d)$ *are given. Then, under the conditions of Theorem 3.1, there exists a unique weak solution to* (1).

Proof Under the conditions, from the definitions of a_0 and l_0, we can show that a_0 is a continuous symmetric bilinear form on $X \times X$ and l_0 is a continuous linear form on X.

We set $\tilde{u} := u - g$. Then u is a weak solution to (1) if and only if

$$\tilde{u} \in V, \quad a_0(\tilde{u}, v) = l_0(v) + a_0(g, v) \quad (v \in V). \tag{3}$$

From the Lax–Milgram lemma [1] and Theorem 3.1, there exists a unique $\tilde{u}$ which satisfies (3). Hence the unique existence of the weak solution has been proved. □

A variational principle for the above symmetric Lax–Milgram type problem is also well known. The weak solution $u^* \in V(g)$ is a unique minimizer of the following energy:

$$u^* = \underset{u \in V(g)}{\arg\min}\, E(u), \tag{4}$$

where

$$E(u) := \frac{1}{2} a_0(u, u) - l_0(u). \tag{5}$$

In Sect. 5, we consider a finite element approximation for our model (1). So-called Céa's lemma [1] implies the following error estimate. We define

$$a^* := \sup_{v,w\in V} \frac{a_0(v,w)}{\|v\|_X\|w\|_X} < \infty.$$

Proposition 3.3 *Under the assumptions of Theorem 3.2, we suppose that V_h is a closed subspace of V. Then there uniquely exists u_h such that*

$$u_h \in V_h(g) := V_h + g, \quad a_0(u_h, v_h) = l_0(v_h) \quad (v_h \in V_h). \tag{6}$$

Furthermore, it satisfies

$$\|u - u_h\|_X \le \frac{a^*}{a_*} \inf_{v_h\in V_h(g)} \|u - v_h\|_X.$$

The problem (6) corresponds to a finite element scheme. If V_h is a space of piecewise linear element (P1 element) on a regular triangular mesh, it is known that $\inf_{v_h\in V_h(g)} \|u - v_h\|_X = O(h)$ as the mesh size h tends to 0 under suitable regularity for u and the triangular mesh [1].

4 Alternating Iteration Method

We remark that $u^* = (u_1^*, u_2^*) \in V(g)$ is a weak solution to (1) if and only if

$$a_1(u_1^*, v_1) + a_\Gamma(u_1^*, v_1) = a_\Gamma(u_2^*, v_1) + l_1(v_1) \quad ({}^\forall v_1 \in V_1), \tag{7}$$

$$a_2(u_2^*, v_2) + a_\Gamma(u_2^*, v_2) = a_\Gamma(u_1^*, v_2) + l_2(v_2) \quad ({}^\forall v_2 \in V_2). \tag{8}$$

We consider the following alternating method.

Gauss–Seidel type scheme

For given $u_2^0 \in V_2(g)$, and for without m = 0, because u_2^0 is given and only $U_1^m, u_2^m m >= 1$ are found by equations (9) and (10) find $u^m = (u_1^m, u_2^m) \in V(g)$ such that

$$a_1(u_1^m, v_1) + a_\Gamma(u_1^m, v_1) = a_\Gamma(u_2^{m-1}, v_1) + l_1(v_1) \quad ({}^\forall v_1 \in V_1), \tag{9}$$

$$a_2(u_2^m, v_2) + a_\Gamma(u_2^m, v_2) = a_\Gamma(u_1^m, v_2) + l_2(v_2) \quad ({}^\forall v_2 \in V_2). \tag{10}$$

We call the above alternating iteration method "Gauss–Seidel type" by analogy with an iterative solver for linear systems. The unique solvability of each u_i^m is clear from Körn's second inequality.

The following theorem tells us that the Gauss–Seidel type scheme defines a sequence $\{u^m\}$ which monotonically decreases the total energy $E(u)$ defined in (5).

Theorem 4.1 *The obtained sequence* $\{u^m = (u_1^m, u_2^m)\}_m$ *by the Gauss–Seidel type scheme satisfies the following energy decay property:*

$$E((u_1^{m-1}, u_2^{m-1})) \geq E((u_1^m, u_2^{m-1})) \geq E((u_1^m, u_2^m)) \quad (m = 1, 2, \ldots). \tag{11}$$

Proof For simplicity, we denote $(u_1^m, u_2^n) \in V(g)$ by $u^{m,n}$. The second inequality is shown as follows. Setting $v = (v_1, v_2) := u^{m,m-1} - u^{m,m} = (0, u_2^{m-1} - u_2^m)$, we have

$$\begin{aligned}
&E(u^{m,m-1}) - E(u^{m,m})\\
&= \frac{1}{2}a_0(u^{m,m-1} + u^{m,m}, u^{m,m-1} - u^{m,m}) - l_0(u^{m,m-1} - u^{m,m})\\
&= \frac{1}{2}a_0(v, v) + a_0(u^{m,m}, v) - l_0(v)\\
&\geq a_0(u^{m,m}, v) - l_0(v)\\
&= a_1(u_1^m, v_1) + a_2(u_2^m, v_2) + a_\Gamma(u_2^m - u_1^m, v_2 - v_1) - l_2(v_2)\\
&= 0,
\end{aligned}$$

where we have used $v_1 = 0$ and (10) for the last equality. The first inequality in (11) is shown in the same way. □

Since the weak solution u^* is the minimizer of the total energy E as written in (4), the sequence generated by the Gauss–Seidel type scheme is expected to approximate u^*. We will study it numerically in the next section.

5 Numerical Solution

First we recall the assumption $\mathcal{H}^{d-1}(\partial_D\Omega) > 0$. In this section, we only deal with the case $\partial_D\Omega \subset \partial\Omega_1$ and $\zeta(x) > 0$ on Γ.

5.1 A Matrix Representation of the Monolithic Formulation

We briefly review a matrix representation of the monolithic formulation. Let $\Lambda^{(i)}$ be an index for finite element basis function and be decomposed as $\Lambda_I^{(i)} \cup \Lambda_B^{(i)}$, corresponding to nodes in the subdomain $\overline{\Omega_i} \setminus \Gamma$ and on the common boundary Γ. We define the following stiffness matrices in each subdomain Ω_i using the bilinear forms $\{a_i(\cdot, \cdot)\}_{i=1,2}$ defined in Sect. 2:

$$[A^{(i)}_{\mu\nu}]_{kl} = a_i(\varphi^{(i)}_l, \varphi^{(i)}_k) \qquad (k \in \Lambda^{(i)}_\mu, l \in \Lambda^{(i)}_\nu, \{\mu, \nu\} \in \{I, B\}),$$
$$[M^{(ij)}]_{kl} = a_\Gamma(\varphi^{(j)}_l, \varphi^{(i)}_k) \qquad (i, j \in \{1, 2\},\ k \in \Lambda^{(i)}_B, l \in \Lambda^{(j)}_B).$$

Here combination of four mass matrices $\{M^{(ij)}\}$ provides a matrix representation of the bilinear form $a_\Gamma([\cdot], [\cdot])$. A matrix representation of the monolithic formulation reads

$$\begin{bmatrix} A^{(1)}_{II} & A^{(1)}_{IB} & & \\ A^{(1)}_{BI} & A^{(1)}_{BB} + M^{(11)} & -M^{(12)} & \\ & -M^{(21)} & A^{(2)}_{BB} + M^{(22)} & A^{(2)}_{BI} \\ & & A^{(2)}_{IB} & A^{(2)}_{II} \end{bmatrix} \begin{bmatrix} u_{1,I} \\ u_{1,B} \\ u_{2,B} \\ u_{2,I} \end{bmatrix} = \begin{bmatrix} f_{1,I} \\ f_{1,B} \\ f_{2,B} \\ f_{2,I} \end{bmatrix}, \tag{12}$$

where the right hand side consists of the body force f and inhomogenous Dirichlet and Neumann data g and q.

Remark 5.1 The monolithic method can be computed by LDU-factorization with any symmetric permutation because the stiffness matrix is positive definite thanks to the coercivity of the weak form with $\mathcal{H}^{d-1}(\partial_D \Omega) > 0$ (Theorem 3.1, Proposition 3.3).

5.2 Alternating Iterative Method in Discrete Form

In the following, we suppose that the nodal points and the surface meshes of the mesh decomposition of domains Ω_1 and Ω_2 coincide on the interface Γ, which leads to

$$M = M^{(11)} = M^{(12)} = M^{(21)} = M^{(22)}.$$

Let us define an inner product of vector $u_{i,\mu}$ and $v_{i,\mu}$ as $(u_{i,\mu}, v_{i,\mu}) := \sum_{k \in \Lambda^{(i)}_\mu} (u_\mu)_k (v_\mu)_k$ for $i = 1, 2$ and $\mu \in \{I, B\}$, and denote the standard ℓ^2-norm by $\|u_{i,\mu}\| = (u_{i,\mu}, u_{i,\mu})^{1/2}$. Since $\zeta(x) > 0$ on Γ, the mass matrix M is positive definite and $(M u_{i,B}, v_{i,B})$ becomes an inner product with weight M. Hence, we denote a norm with the weight M by $\|u_{i,B}\|_M = (M\, u_{i,B}, u_{i,B})^{\frac{1}{2}}$. Then there exist $\beta_1 > 0$ and $\beta_2 > 0$ such that $\beta_1 \|u_{i,B}\|^2_M \le \|u_{i,B}\|^2 \le \beta_2 \|u_{i,B}\|^2_M$ holds for any $u_{i,B}$.

We prepare two matrices in each subdomain to describe the linear system in a simpler way,

$$\mathcal{A}^{(i)} := \begin{bmatrix} A^{(i)}_{II} & A^{(i)}_{IB} \\ A^{(1)}_{BI} & A^{(i)}_{BB} \end{bmatrix}, \quad \widetilde{\mathcal{A}}^{(i)} := \begin{bmatrix} A^{(i)}_{II} & A^{(i)}_{IB} \\ A^{(i)}_{BI} & A^{(i)}_{BB} + M \end{bmatrix}.$$

Lemma 5.2 *There exists* $\alpha_1 > 0$ *such that*

$$(\mathcal{A}^{(1)}[v_{1,I}\ \ v_{1,B}]^T, [v_{1,I}\ \ v_{1,B}]^T) \ge \alpha_1(\|v_{1,I}\|^2 + \|v_{1,B}\|^2) \ge \alpha_1 \beta_1 \|v_{1,B}\|^2_M$$

holds for any vector $[v_{1,I}\ \ v_{1,B}]^T$.

Proof Since $\mathcal{A}^{(1)}$ in Ω_1 is positive definite due to the Dirichlet boundary $\partial_D\Omega \subset \partial\Omega_1$, there exists $\alpha_1 > 0$ and the first inequality holds. The second one is clear from the definition of β_1. □

The Gauss–Seidel type iteration defined in (9) and (10) is written in the following matrix presentation.

Algorithm 1 (*Gauss–Seidel type iteration*) Let $u_{2,B}^0$ be an initial guess of u_2 on Γ. From obtained data $u_{2,B}^{m-1}$ of $m-1$-step, approximate solution $[u_{1,I}^m\ u_{1,B}^m]^T$ and $[u_{2,I}^m\ u_{2,B}^m]^T$ are generated by solving following two problems successively,

$$\widetilde{\mathcal{A}}^{(1)}\begin{bmatrix} u_{1,I}^m \\ u_{1,B}^m \end{bmatrix} = \begin{bmatrix} f_{1,I} \\ f_{1,B} + M\, u_{2,B}^{m-1} \end{bmatrix} \text{ and then } \widetilde{\mathcal{A}}^{(2)}\begin{bmatrix} u_{2,I}^m \\ u_{2,B}^m \end{bmatrix} = \begin{bmatrix} f_{2,I} \\ f_{2,B} + M\, u_{1,B}^m \end{bmatrix}.$$

Let $[u_{1,I}^m\ u_{1,B}^m]^T$ and $[u_{2,I}^m\ u_{2,B}^m]^T$ $(m = 1, 2, \ldots)$ be the solution of the Gauss–Seidel type iteration and let $[u_{1,I}^*\ u_{1,B}^*]^T$ and $[u_{2,I}^*\ u_{2,B}^*]^T$ be the one of the monolithic system (12). We define the error between them by

$$[e_{i,I}^m\ e_{i,B}^m]^T := [u_{i,I}^m\ u_{i,B}^m]^T - [u_{i,I}^*\ u_{i,B}^*]^T \quad (i = 1, 2). \tag{13}$$

We have the following convergence estimates for the Gauss–Seidel type iteration.

Lemma 5.3 *The error on the boundary admits the following estimates:*

$$\|e_{1,B}^m\|_M \le r\, \|e_{2,B}^{m-1}\|_M, \quad \|e_{2,B}^m\|_M \le \|e_{1,B}^m\|_M \quad (m = 1, 2, \ldots),$$

where $r := 1/(1 + \alpha_1\beta_1) < 1$.

Proof From the definition of the errors (13), they satisfy the following linear systems:

$$\widetilde{\mathcal{A}}^{(1)}[e_{1,I}^m\ e_{1,B}^m]^T = [0\ \ M\, e_{2,B}^{m-1}]^T, \quad \widetilde{\mathcal{A}}^{(2)}[e_{2,I}^m\ e_{2,B}^m]^T = [0\ \ M\, e_{1,B}^m]^T. \tag{14}$$

Taking inner product of $[e_{1,I}^m\ e_{1,B}^m]^T$ with the left equation of (14), and of $[e_{2,I}^m\ e_{2,B}^m]^T$ with the right, we obtain

$$\left(\mathcal{A}^{(1)}[e_{1,I}^m\ e_{1,B}^m]^T, [e_{1,I}^m\ e_{1,B}^m]^T\right) + \|e_{1,B}^m\|_M^2 = (M\, e_{2,B}^{m-1}, e_{1,B}^m) \le \|e_{2,B}^{m-1}\|_M \|e_{1,B}^m\|_M, \tag{15}$$

$$\left(\mathcal{A}^{(2)}[e_{2,I}^m\ e_{2,B}^m]^T, [e_{2,I}^m\ e_{2,B}^m]^T\right) + \|e_{2,B}^m\|_M^2 = (M\, e_{1,B}^m, e_{2,B}^m) \le \|e_{1,B}^m\|_M \|e_{2,B}^m\|_M. \tag{16}$$

From Lemma 5.2 and (15), we have

$$(1 + \alpha_1\beta_1)\|e_{1,B}^m\|_M^2 \le \|e_{2,B}^{m-1}\|_M \|e_{1,B}^m\|_M.$$

This gives the first inequality. The second inequality is obtained from (16) with positive semi-definiteness of $\mathcal{A}^{(2)}$. □

Theorem 5.4 *There exists $C > 0$ such that the following error estimate holds:*

$$\sqrt{\|e_{i,I}^m\|^2 + \|e_{i,B}^m\|^2} \leq C r^{m+i-2} \|e_{2,B}^0\|_M \quad (i = 1, 2,\ m = 1, 2, \ldots),$$

where $r := 1/(1 + \alpha_1 \beta_1) < 1$.

Proof From Lemma 5.3, we have

$$\|e_{2,B}^m\|_M \leq \|e_{1,B}^m\|_M \leq r\, \|e_{2,B}^{m-1}\|_M \quad (m = 1, 2, \ldots).$$

These inequalities imply

$$\|e_{i,B}^m\|_M \leq r^m\, \|e_{2,B}^0\|_M \quad (i = 1, 2,\ m = 1, 2, \ldots). \tag{17}$$

Since the matrices $\widetilde{\mathcal{A}}^{(i)}$ $(i = 1, 2)$ are invertible, from (14), we obtain

$$\sqrt{\|e_{1,I}^m\|^2 + \|e_{1,B}^m\|^2} \leq \left\|(\widetilde{\mathcal{A}}^{(1)})^{-1}\right\|\ \|M e_{2,B}^{m-1}\| = \left\|(\widetilde{\mathcal{A}}^{(1)})^{-1}\right\|\ \|M^{1/2} e_{2,B}^{m-1}\|_M\,,$$
$$\sqrt{\|e_{2,I}^m\|^2 + \|e_{2,B}^m\|^2} \leq \left\|(\widetilde{\mathcal{A}}^{(2)})^{-1}\right\|\ \|M e_{1,B}^{m}\| = \left\|(\widetilde{\mathcal{A}}^{(2)})^{-1}\right\|\ \|M^{1/2} e_{1,B}^{m}\|_M\,.$$

Together with the estimate (17) and with the fact that M is positive definite, there exists a $C > 0$ such that the assertion of the theorem holds. □

Remark 5.5 The Gauss–Seidel type iteration is straightforwardly extended to SOR type iteration by introducing a relaxation parameter.

5.3 *Numerical Results*

Now we show numerical verification on convergence of Gauss–Seidel type iteration using a manufactured solution,

$$\begin{bmatrix} u_1 \\ u_2 \\ u_3 \end{bmatrix} = \begin{bmatrix} \sin((\pi/8)\,x_1) \times \cos((\pi/8)\,x_2) \times \sin((\pi/16)\,x_3) \\ \cos((\pi/16)\,x_1) \times \sin((\pi/8)\,x_2) \times \cos((\pi/8)\,x_3) \\ \cos((\pi/8)\,x_1) \times \sin((\pi/16)\,x_2) \times \sin((\pi/8)\,x_3) \end{bmatrix}$$

in $\Omega_1 = (0, 4) \times (0, 2) \times (0, 4)$, $\Omega_2 = (0, 4) \times (2, 4) \times (0, 4)$, and $\partial_D \Omega = \{(x, y, z)\,;\ 0 < x < 4, y = 0, 0 < z < 4\}$ with corresponding inhomogeneous Dirichlet data $g(x)$, the load $f(x)$ on Ω, and the surface traction $q(x)$ on $\partial_N \Omega$. For simplicity, we put $\zeta(x) \equiv 1$. Figure 3 shows relative errors of finite element solution discretized with P1 element solved by the monolithic formulation, which ensures the first-order approximation error of the solution, $O(h)$ with mesh size h. Convergence

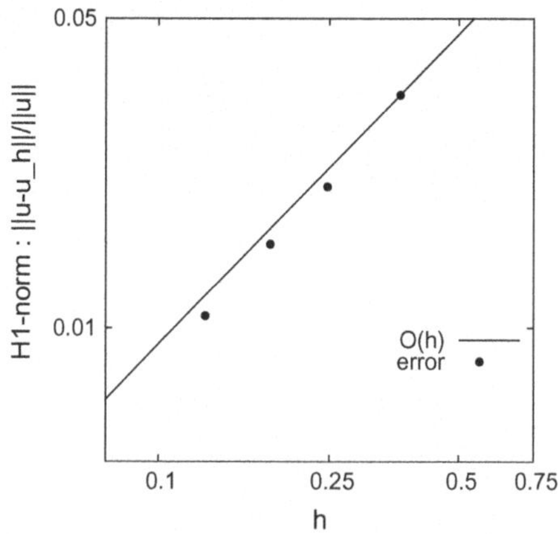

Fig. 3 Relative errors of P1 finite element solution computed by monolithic formulation

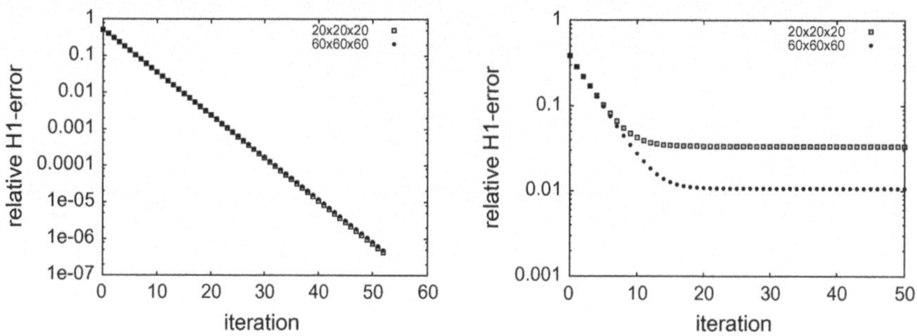

Fig. 4 Convergence history of Gauss–Seidel type iteration

of Gauss–Seidel type iteration to the solution by the monolithic formulation with the relative error measured by $\|\cdot\|_{H^1(\Omega)}$ is shown in the left of Fig. 4, and relative error to the manufactured solution in the right of Fig. 4. Here mesh subdivisions with $h_{\max} = 0.36551$ in $20 \times 20 \times 20$, and $h_{\max} = 0.12878$ in $60 \times 60 \times 60$ are used. We can see the convergence does not depend on the mesh size, and that the same relative error as one by monolithic formulation to the manufactured solution is obtained after certain iterations, though Gauss–Seidel type iteration continues to converge.

5.4 Computational Efficiency

We used FreeFem++ software package and Dissection sparse direct solver [6] to obtain finite element solution. Table 1 shows computational time of the direct solver for the monolithic formulation and the Gauss–Seidel type iteration, using Intel Core processor i7-6770HQ with four cores running at 2.60GHz. LDU-factorization is

Table 1 Elapsed time of monolithic and Gauss–Seidel methods by FreeFem++ and Dissection solver with four cores. CPU time for factorization are also shown within parentheses

Mesh	$h_{\max}$	Solver	# Unknowns	# Iteration	Time (s)			
					Factorization		Iteration	Total
20 × 20 × 20	0.36551	Monolithic	30,870	–	1.364	(3.147)	–	1.364
		Gauss–Seidel	14,742+16,128	12	1.339	(2.752)	0.245	1.585
30 × 30 × 30	0.24729	Monolithic	97,743	–	6.901	(18.588)	–	6.901
		Gauss–Seidel	46,965+50,778	15	5.082	(12.425)	1.286	6.368
60 × 60 × 60	0.12783	Monolithic	748,470	–	174.123	(603.567)	–	174.123
		Gauss–Seidel	368,928+379,542	20	118.973	(316.482)	25.559	144.532

performed before starting iteration and forward/backward substitutions are repeated because the stiffness matrix in each subdomain does not change during the iteration. Since computational complexity of LDU-factorization of sparse matrix with P1 finite element is more than $O(N^2)$ with number of unknowns N, factorization cost for sub-matrices in Gauss–Seidel type is less than half, $2 \times (N/2)^2/N^2 = 1/2$ of the one for monolithic formulation. We can observe shorter CPU time that is total time spent by all cores for factorization of Gauss–Seidel type, which is shown as number in parentheses, when the problem size is enough large. In Dissection solver, numerical factorization is fully parallelized but there exist some sequential processing parts, e.g., fill-in analysis of the sparse matrix, which masks speed-up of elapsed time for the factorization in Gauss–Seidel type. We also observe that elapsed time of iteration in Gauss–Seidel type with selected iterating number to obtain appropriately approximate solution is less than time of the factorization. Hence if we can find a reasonable criteria to stop iteration, the Gauss–Seidel type iteration becomes more efficient than monolithic formulation.

6 Conclusion

We considered a stationary deformation problem for glued elastic bodies and have established solvability of its weak formulation. We proposed a kind of alternating iteration scheme to approximate the problem and showed that the scheme has a nature of alternating minimizing algorithm with respect to the total energy.

We proved the convergence for the Gauss–Seidel type iteration in a rate of $O(r^m)$ with $r \in (0, 1)$ in discrete setting. Computational efficiency of the monolithic and alternating iterative algorithms have been verified with a three-dimensional problem. The alternating iterative method requires smaller computational resource than the monolithic method and also it has an advantage in computation time when the degree of freedom is sufficiently large.

Acknowledgments The authors are grateful to Prof. Frédéric Hecht for his useful comments on numerical computation with FreeFem++. This work is supported by JSPS KAKENHI Grant Number JP16H03946 and JP17H02857.

References

1. Ciarlet, P.G.: The Finite Element Method for Elliptic Problems. North-Holland, Amsterdam (1978)
2. Duvaut, G., Lions, J.L.: Inequalities in Mechanics and Physics. Springer, Berlin (1976)
3. Frémond, M.: Non-Smooth Thermomechanics. Springer, Berlin (2002)
4. Roubíček, T., Scardia, L., Zanini, C.: Quasistatic delamination problem. Cont. Mech. Themodyn. **21**, 223–235 (2009). https://doi.org/10.1007/s00161-009-0106-4
5. Scala, R.: Limit of viscous dynamic processes in delamination as the viscosity and inertia vanish. ESAIM: COCV, **23**, 593–625 (2017). https://doi.org/10.1051/cocv/2016006

6. Suzuki, A., Roux, F.-X.: A dissection solver with kernel detection for symmetric finite element matrices on shared memory computers. Int. J. Numer. Methods Eng. **100**, 136–164 (2014). https://doi.org/10.1007/s00161-009-0106-4
7. Yoneda, T.: Finite element analysis of a vibration-delamination model. Master Thesis, Graduate School of Natural Science and Technology, Kanazawa University, 48 p (2018). (in Japanese)

Shape Optimization and Inverse Problems

Shape Differentiability of Lagrangians and Application to Overdetermined Problems

Victor A. Kovtunenko and Kohji Ohtsuka

Abstract A class of geometry-dependent Lagrangians is investigated in a functional analysis framework with respect to the property of shape differentiability. General results are presented due to Delfour–Zolésio who adopted to shape optimization an abstract theorem of Correa–Seeger on the directional differentiability. A crucial point concerns the bijective property of function spaces as well as their feasible sets that must be preserved under a kinematic flow of geometry. The shape differentiability result is applied to overdetermined free-boundary and inverse problems expressed by least-square solutions. The theory is supported by explicit formulas obtained for calculation of the shape derivative.

Keywords Shape derivative · Lagrangian · Correa–Seeger theorem · Delfour–Zolésio theorem · State-constrained shape optimization · Free-boundary · Inverse problem

1 Introduction

In this work, we establish the shape differentiability of Lagrangians expressed by solutions of saddle-point problems stated over geometry-dependent function spaces. As examples, overdetermined problems of free-boundary and inverse types in the form of least-square minimization subject to state-constraints are considered.

V. A. Kovtunenko (✉)
Institute for Mathematics and Scientific Computing, Karl-Franzens University of Graz, NAWI Graz, Heinrichstr. 36, Graz 8010, Austria
e-mail: victor.kovtunenko@uni-graz.at

Lavrentyev Institute of Hydrodynamics, Siberian Division of the Russian Academy of Sciences, 630090 Novosibirsk, Russia

K. Ohtsuka
Faculty of Information Design and Sociology, Hiroshima Kokusai Gakuin University, 6-20-1, Aki-ku, Hiroshima 739-0321, Japan
e-mail: ohtsuka@hkg.ac.jp

H. Itou et al. (eds.), *Mathematical Analysis of Continuum Mechanics and Industrial Applications III*, Mathematics for Industry 34,
https://doi.org/10.1007/978-981-15-6062-0_7

Our motivation came from applications of shape and topology optimization methods to variational problems in mechanics and other engineering sciences, see [15, 27]. Especially aimed at fracture and earthquakes, the variational concept of nonlinear crack problems subject to non-penetration was developed in [10–13] and other works by the authors. Widely used here Griffith fracture criterion implies the shape derivative of the strain energy with respect to a crack perturbation, which was calculated, e.g., in [14, 19] for rectilinear cracks. For the related J-integrals and generalized J-integrals, we refer to [2, 24, 25]. Based on the velocity method, see, e.g., [9], an explicit formula of the shape derivative is provided by the bijection property of feasible sets which, however, fails for curvilinear cracks. This drawback was remedied in [22, 26] by a Gamma-type convergence, and in [16, 17] based on a Lagrangian setting of the perturbation problem.

The minimization problems subject to unilateral constraints can be reset equivalently as saddle-point problems using Lagrangian formalism. Within this concept, in [20] the Stokes problem, and in [6] the Brinkman problem subject to the divergence-free constraint were analyzed with respect to its shape differentiability, provided by the bijection property of function spaces under mixed Dirichlet–Neumann boundary conditions, and by a counter-example of non-bijection under no-slip conditions. In the present work, we get a shape derivative for the least-square minimization subject to state-constraints which approximates overdetermined boundary value problems. Examples are the Bernoulli free-boundary problem, see, e.g., [8, 23], and the inverse problem of parameter identification from boundary measurements, see [1, 3, 7, 18, 21].

In a functional analysis framework, in Sect. 2 we recall the Correa–Seeger theorem on directional differentiability [4], the Delfour–Zolésio theorem on differentiability with respect to a parameter [5, Chap. 10, Theorem 5.1], and in Sect. 3 we establish the shape differentiability of geometry-dependent Lagrangians under the bijection property of underlying function spaces and feasible sets. An application to the state-constrained shape optimization is given in Sect. 4.

2 Directional Differentiability of Lagrangians

In a locally convex space O, Hausdorff topological spaces V and $H^\star$ with nonempty subsets $K \subseteq V$ and $K^\star \subseteq H^\star$, we consider an abstract *Lagrangian function*

$$\mathscr{L}(\phi, w, p) : \ O \times V \times H^\star \mapsto \overline{\mathbb{R}} \tag{1}$$

obeying optimal values $l_\phi, l^\phi \in \overline{\mathbb{R}}$ such that

$$l_\phi := \sup_{p \in K^\star} \inf_{w \in K} \mathscr{L}(\phi, w, p) \leq \inf_{w \in K} \sup_{p \in K^\star} \mathscr{L}(\phi, w, p) =: l^\phi. \tag{2}$$

For given $\phi, d \in O$, the *solution sets* corresponding to (2) are determined by

$$K_\phi = \{u \in K \mid \sup_{p \in K^\star} \mathscr{L}(\phi, u, p) = l^\phi\}, \quad K_\phi^\star = \{\lambda \in K^\star \mid \inf_{w \in K} \mathscr{L}(\phi, w, \lambda) = l_\phi\} \tag{3}$$

yielding multi-valued functions by the mean of

$$\mathbb{R}_+ \rightrightarrows K, \ s \rightrightarrows K_{\phi+sd}, \quad \mathbb{R}_+ \rightrightarrows K^\star, \ s \rightrightarrows K^\star_{\phi+sd}. \tag{4}$$

Assume that there exists $\delta > 0$ such that the following properties are fulfilled:

(P0) $K_{\phi+sd}$ and $K^\star_{\phi+sd}$ are *sequentially semicontinuous* at 0, that is, for every convergent sequence $s_k \to 0^+$ as $k \to \infty$ there exists an accumulation point

$$u_\phi \in K_\phi, \quad \lambda_\phi \in K_\phi^\star \tag{5}$$

and a sequence $(u_{\phi+s_kd}, \lambda_{\phi+s_kd})$ accumulating at (u_ϕ, λ_ϕ) such that for all $k \in \mathbb{N}$ sufficiently large

$$u_{\phi+s_kd} \in K_{\phi+s_kd}, \quad \lambda_{\phi+s_kd} \in K^\star_{\phi+s_kd}; \tag{6}$$

(P1) for every $(w, p) \in K \times K^\star$, the function $\mathbb{R}_+ \mapsto \overline{\mathbb{R}}$, $s \mapsto \mathscr{L}(\phi + sd, w, p)$ is finite and continuous in $[0, \delta)$;

(P2) for all $u \in K_\phi$ the function

$$\mathbb{R}_+ \times K^\star \mapsto \overline{\mathbb{R}}, \quad (s, p) \mapsto \liminf_{\tau \to 0^+} \frac{\mathscr{L}(\phi+(s+\tau)d, u, p) - \mathscr{L}(\phi+sd, u, p)}{\tau} \tag{7}$$

is finite and upper semicontinuous in $\{0\} \times K_\phi^\star$; for all $\lambda \in K_\phi^\star$ the function

$$\mathbb{R}_+ \times K \mapsto \overline{\mathbb{R}}, \quad (s, w) \mapsto \limsup_{\tau \to 0^+} \frac{\mathscr{L}(\phi+(s+\tau)d, w, \lambda) - \mathscr{L}(\phi+sd, w, \lambda)}{\tau} \tag{8}$$

is finite and lower semicontinuous in $\{0\} \times K_\phi$;

(P3) the *saddle-point property* $l_{\phi+sd} = l^{\phi+sd}$ holds for all $s \in [0, \delta)$.

Theorem 1 (Correa–Seeger) *If properties (P0)–(P3) hold, then a directional derivative exists, which is characterized by*

$$\begin{aligned}\lim_{s_k \to 0^+} \frac{\mathscr{L}(\phi+s_kd, u_{\phi+s_kd}, \lambda_{\phi+s_kd}) - \mathscr{L}(\phi, u_\phi, \lambda_\phi)}{s_k} &= \sup_{\lambda \in K_\phi^\star} \inf_{u \in K_\phi} \left(\lim_{s \to 0^+} \frac{\mathscr{L}(\phi+sd, u, \lambda) - \mathscr{L}(\phi, u, \lambda)}{s} \right) \\ &= \inf_{u \in K_\phi} \sup_{\lambda \in K_\phi^\star} \left(\lim_{s \to 0^+} \frac{\mathscr{L}(\phi+sd, u, \lambda) - \mathscr{L}(\phi, u, \lambda)}{s} \right).\end{aligned} \tag{9}$$

Let us now consider the abstract Lagrangian function $\mathscr{L}(t, w, p)$ with a parameter $t_0 < t < t_1$. Using Theorem 1, we have the directional derivative of $\mathscr{L}(t + s, u_{t+s}, \lambda_{t+s})$ with respect to $s \to 0^+$ by setting $O = (t_0, t_1)$, $\phi = t$, $d = 1$, that is, $\mathscr{L}(\phi + sd, w, p) = \mathscr{L}(t + s, w, p)$, $l_{\phi+sd} = l_{t+s}$, $l^{\phi+sd} = l^{t+s}$, $K_{\phi+sd} = K_{t+s}$,

$K^\star_{\phi+sd} = K^\star_{t+s}, u_{\phi+sd} = u_{t+s}, \lambda_{\phi+sd} = \lambda_{t+s}$ in (1)–(9). The derivative can be achieved as follows:

Assume that $0 < \delta < t_1 - t$ and $(\mathfrak{T}_V, \mathfrak{T}_{H^\star})$-topology in $V \times H^\star$ exist such that

(H1) for all $s \in [0, \delta)$ the set of saddle-points $\mathfrak{S}(t+s) \subseteq K_{t+s} \times K^\star_{t+s}$ is nonempty:

$$\mathfrak{S}(t+s) := \{(u_{t+s}, \lambda_{t+s})| \quad l_{t+s} = \mathscr{L}(t, u_{t+s}, \lambda_{t+s}) = l^{t+s}\}; \tag{10}$$

(H2) for all $(u, \lambda) \in \big(\bigcup_{s\in[0,\delta)} K_{t+s} \times K^\star_t \big) \cup \big(K_t \times \bigcup_{s\in[0,\delta)} K^\star_{t+s} \big)$ there exists the *partial derivative* with respect to parameter t:

$$\tfrac{\partial}{\partial t}\mathscr{L}(t, u, \lambda) = \lim_{s\to 0^+} \tfrac{\mathscr{L}(t+s,u,\lambda)-\mathscr{L}(t,u,\lambda)}{s} \quad \text{(one-sided)}; \tag{11}$$

(H3) as $s \to 0^+$ an accumulation point $u_t \in K_t$ and a sequence $u_{t+s_k} \in K_{t+s_k}$ exist such that

$$u_{t+s_k} \to u_t \quad \text{strongly in } \mathfrak{T}_V - \text{topology} \tag{12}$$

and the lower estimate holds:

$$\liminf_{\tau, s_k\to 0^+} \tfrac{\partial}{\partial t}\mathscr{L}(t+\tau, u_{t+s_k}, \lambda) \geq \tfrac{\partial}{\partial t}\mathscr{L}(t, u_t, \lambda) \quad \forall \lambda \in K^\star_t; \tag{13}$$

(H4) as $s \to 0^+$ an accumulation point $\lambda_t \in K^\star_t$ and a sequence $\lambda_{t+s_k} \in K^\star_{t+s_k}$ exist such that

$$\lambda_{t+s_k} \to \lambda_t \quad \text{strongly in } \mathfrak{T}_{H^\star} - \text{topology} \tag{14}$$

and the upper estimate holds:

$$\limsup_{\tau, s_k\to 0^+} \tfrac{\partial}{\partial t}\mathscr{L}(t+\tau, u, \lambda_{t+s_k}) \leq \tfrac{\partial}{\partial t}\mathscr{L}(t, u, \lambda_t) \quad \forall u \in K_t. \tag{15}$$

Theorem 2 (Delfour–Zolésio) *Under hypotheses (H1)–(H4) there exists a saddle-point (u_t, λ_t) of $\frac{\partial}{\partial t}\mathscr{L}(t, u, \lambda)$ on $K_t \times K^\star_t$ such that*

$$\sup_{\lambda\in K^\star_t} \inf_{u\in K_t} \tfrac{\partial}{\partial t}\mathscr{L}(t, u, \lambda) = \tfrac{\partial}{\partial t}\mathscr{L}(t, u_t, \lambda_t) = \inf_{u\in K_t} \sup_{\lambda\in K^\star_t} \tfrac{\partial}{\partial t}\mathscr{L}(t, u, \lambda) \tag{16}$$

and a derivative with respect to parameter, which is represented by the partial derivative:

$$\lim_{s_k\to 0^+} \tfrac{\mathscr{L}(t+s_k,u_{t+s_k},\lambda_{t+s_k})-\mathscr{L}(t,u_t,\lambda_t)}{s_k} = \tfrac{\partial}{\partial t}\mathscr{L}(t, u_t, \lambda_t) \quad \textit{(one-sided)}. \tag{17}$$

3 Shape Differentiability of Lagrangians

In the following we adapt (10)–(17) for the reason of shape differentiability. For the parameter $t \in (t_0, t_1)$ we associate a *reference geometry*

$$t \mapsto \Omega_t \subset \mathbb{R}^d, \quad d \in \mathbb{N} \tag{18}$$

to a *geometry-dependent Lagrangian* over topological spaces $V(\Omega_t)$ and $H^\star(\Omega_t)$:

$$V(\Omega_t) \times H^\star(\Omega_t) \mapsto \mathbb{R}, \quad (w, p) \mapsto \mathscr{L}(w, p; \Omega_t). \tag{19}$$

The corresponding optimal values are defined by

$$l_t := \sup_{p \in K^\star(\Omega_t)} \inf_{w \in K(\Omega_t)} \mathscr{L}(w, p; \Omega_t) \leq \inf_{w \in K(\Omega_t)} \sup_{p \in K^\star(\Omega_t)} \mathscr{L}(w, p; \Omega_t) =: l^t \tag{20}$$

over feasible sets $K(\Omega_t) \subseteq V(\Omega_t)$ and $K^\star(\Omega_t) \subseteq H^\star(\Omega_t)$, the solution sets are

$$\begin{aligned} K_t &= \{u \in K(\Omega_t) | \sup_{p \in K^\star(\Omega_t)} \mathscr{L}(u, p; \Omega_t) = l^t\}, \\ K_t^\star &= \{\lambda \in K^\star(\Omega_t) | \inf_{w \in K(\Omega_t)} \mathscr{L}(w, \lambda; \Omega_t) = l_t\}. \end{aligned} \tag{21}$$

For a small perturbation parameter $s \in (t_0 - t, t_1 - t)$ and the *perturbed geometry* Ω_{t+s} the *perturbed Lagrangian* (19) reads:

$$V(\Omega_{t+s}) \times H^\star(\Omega_{t+s}), \quad (v, \mu) \mapsto \mathscr{L}(v, \mu; \Omega_{t+s}), \tag{22}$$

the *perturbed subsets* $K(\Omega_{t+s}) \subseteq V(\Omega_{t+s})$ and $K^\star(\Omega_{t+s}) \subseteq H^\star(\Omega_{t+s})$, and

$$\begin{aligned} l_{t+s} = &\sup_{\mu \in K^\star(\Omega_{t+s})} \inf_{v \in K(\Omega_{t+s})} \mathscr{L}(v, \mu; \Omega_{t+s}) \\ &\leq \inf_{v \in K(\Omega_{t+s})} \sup_{\mu \in K^\star(\Omega_{t+s})} \mathscr{L}(v, \mu; \Omega_{t+s}) = l^{t+s}. \end{aligned} \tag{23}$$

Definition 1 For *saddle-points* $(u_t, \lambda_t) \in \mathfrak{S}(t) \subseteq K_t \times K_t^\star$ and the *optimal value function* $\ell : (t_0, t_1) \mapsto \overline{\mathbb{R}}$, $\ell(t) = l_t$ written over $\mathfrak{S}(t)$ such that according to (10)

$$\ell(t) = l_t = \mathscr{L}(u_t, \lambda_t; \Omega_t) = l^t, \tag{24}$$

the *shape derivative* of $\mathscr{L}$ is called the following limit (if it exists):

$$\partial_t \mathscr{L}(u_t, \lambda_t; \Omega_t) = \partial \ell(t) := \lim_{s \to 0^+} \tfrac{\ell(t+s) - \ell(t)}{s} \quad \text{(one-sided)}. \tag{25}$$

The principal disadvantage is that Theorem 2 is not applicable to prove the limit in (25) because the function spaces in (19) depend itself on the parameter. By this reason we transform the problem to a fixed geometry. For fixed $t \in (t_0, t_1)$ let

$$[s \mapsto \phi_s], [s \mapsto \phi_s^{-1}] \in C^1([t_0 - t, t_1 - t]; W^{1,\infty}_{\mathrm{loc}}(\mathbb{R}^d; \mathbb{R}^d)) \tag{26}$$

associate the *coordinate transformation* $y = \phi_s(x)$ and its *inverse* $x = \phi_s^{-1}(y)$:

$$(\phi_s^{-1} \circ \phi_s)(x) = x, \quad (\phi_s \circ \phi_s^{-1})(y) = y \tag{27}$$

such that the shape perturbation

$$\Omega_{t+s} = \{y \in \mathbb{R}^d | \quad y = \phi_s(x), \ x \in \Omega_t\} \tag{28}$$

builds the diffeomorphism

$$\phi_s : \Omega_t \mapsto \Omega_{t+s}, \ x \mapsto y, \quad \phi_s^{-1} : \Omega_{t+s} \mapsto \Omega_t, \ y \mapsto x. \tag{29}$$

For example, given a *stationary velocity* $\Lambda(x) \in W^{1,\infty}_{\mathrm{loc}}(\mathbb{R}^d; \mathbb{R}^d)$ establishes the flow (26) determined by the solution of the autonomous ODE system:

$$\begin{cases} \frac{d}{ds}\phi_s = \Lambda(\phi_s), \ s \neq 0, \\ \phi_s = x, \ s = 0, \end{cases} \quad \begin{cases} \frac{d}{ds}\phi_s^{-1} = -\Lambda(\phi_s^{-1}), \ s \neq 0, \\ \phi_s^{-1} = y, \ s = 0 \end{cases} \tag{30}$$

and forms a semigroup of transformations. Generally, the *kinematic velocity* is time-dependent $\Lambda \in C([t_0, t_1]; W^{1,\infty}_{\mathrm{loc}}(\mathbb{R}^d; \mathbb{R}^d))$ defined from (26) as

$$\Lambda(t + s, y) := \tfrac{d}{ds}\phi_s(\phi_s^{-1}(y)). \tag{31}$$

Assume that

(A0) the map $[(v, \mu) \mapsto (v \circ \phi_s, \mu \circ \phi_s)]$ is bijective between the function spaces

$$V(\Omega_{t+s}) \mapsto V(\Omega_t), \quad H^\star(\Omega_{t+s}) \mapsto H^\star(\Omega_t) \tag{32}$$

and between the feasible sets

$$K(\Omega_{t+s}) \mapsto K(\Omega_t), \quad K^\star(\Omega_{t+s}) \mapsto K^\star(\Omega_t). \tag{33}$$

The assumption (32) determines well the *transformed perturbed Lagrangian*:

$$(t_0 - t, t_1 - t) \times V(\Omega_t) \times H^\star(\Omega_t), \quad (s, w, p) \mapsto \tilde{\mathscr{L}}(s, w, p; \Omega_t) \tag{34}$$

in such a way that for arbitrary $(v, \mu) \in V(\Omega_{t+s}) \times H^\star(\Omega_{t+s})$ it holds

$$\tilde{\mathscr{L}}(s, v \circ \phi_s, \mu \circ \phi_s; \Omega_t) = \mathscr{L}(v, \mu; \Omega_{t+s}). \tag{35}$$

We note that (33) may fail for feasible sets involving integral and gradient operators which are generally not preserved under a velocity-induced geometry flow.

Let us put for $s \in (t_0 - t, t_1 - t)$ the optimal values for $\tilde{\mathscr{L}}$:

$$\tilde{l}_{t+s} := \sup_{p \in K^\star(\Omega_t)} \inf_{w \in K(\Omega_t)} \tilde{\mathscr{L}}(s, w, p; \Omega_t) \\ \le \inf_{w \in K(\Omega_t)} \sup_{p \in K^\star(\Omega_t)} \tilde{\mathscr{L}}(s, w, p; \Omega_t) =: \tilde{l}^{t+s} \tag{36}$$

and the corresponding solution sets

$$\begin{aligned} \tilde{K}_{t+s} &= \{u \in K(\Omega_t) | \sup_{p \in K^\star(\Omega_t)} \tilde{\mathscr{L}}(s, u, p; \Omega_t) = \tilde{l}^{t+s}\}, \\ \tilde{K}^\star_{t+s} &= \{\lambda \in K^\star(\Omega_t) | \inf_{w \in K(\Omega_t)} \tilde{\mathscr{L}}(s, w, \lambda; \Omega_t) = \tilde{l}_{t+s}\}. \end{aligned} \tag{37}$$

Definition 2 For saddle-points $(\tilde{u}_{t+s}, \tilde{\lambda}_{t+s}) \in \tilde{\mathfrak{S}}(t+s) \subseteq \tilde{K}_{t+s} \times \tilde{K}^\star_{t+s}$ from the set

$$\tilde{\mathfrak{S}}(t+s) := \{(\tilde{u}_{t+s}, \tilde{\lambda}_{t+s}) | \ \tilde{l}_{t+s} = \tilde{\mathscr{L}}(s, \tilde{u}_{t+s}, \tilde{\lambda}_{t+s}; \Omega_t) = \tilde{l}^{t+s}\} \tag{38}$$

and the *optimal value function* $\tilde{\ell} : (t_0 - t, t_1 - t) \mapsto \overline{\mathbb{R}}$, $\tilde{\ell}(s; t) = \tilde{l}_{t+s}$ written over $\tilde{\mathfrak{S}}(t+s)$ according to (36), the *shape derivative* of $\mathscr{L}$ is called the following limit (if it exists):

$$\partial_s \tilde{\ell}(0; t) := \lim_{s \to 0^+} \tfrac{\tilde{\ell}(s;t) - \tilde{\ell}(0;t)}{s} \quad \text{(one-sided)}. \tag{39}$$

Now Theorem 2 is applicable to prove the limit in (39), when we reformulate the hypotheses (H1)–(H4) as follows. For fixed $t \in (t_0, t_1)$ we assume that there exist $\delta \in (0, t_1 - t)$ and $(\mathfrak{T}_V, \mathfrak{T}_{H^\star})$-topology such that

(A1) for all $s \in [0, \delta)$ the set of saddle-points $\tilde{\mathfrak{S}}(t+s)$ given in (38) is nonempty;

(A2) for all $(u, \lambda) \in \big(\bigcup_{s \in [0,\delta)} \tilde{K}_{t+s} \times \tilde{K}^\star_t \big) \cup \big(\tilde{K}_t \times \bigcup_{s \in [0,\delta)} \tilde{K}^\star_{t+s} \big)$ and $\tau \in [0, \delta)$ there exists the *partial derivative* at $s = \tau$:

$$\frac{\partial}{\partial s} \tilde{\mathscr{L}}(\tau, u, \lambda; \Omega_t) := \lim_{s \to 0^+} \tfrac{\tilde{\mathscr{L}}(\tau + s, u, \lambda; \Omega_t) - \tilde{\mathscr{L}}(\tau, u, \lambda; \Omega_t)}{s} \quad \text{(one-sided)}; \tag{40}$$

(A3) as $s \to 0^+$ an accumulation point $\tilde{u}_t \in \tilde{K}_t$ and a sequence $\tilde{u}_{t+s_k} \in \tilde{K}_{t+s_k}$ exist such that

$$\tilde{u}_{t+s_k} \to \tilde{u}_t \quad \text{strongly in } \mathfrak{T}_V - \text{topology} \tag{41}$$

and the lower estimate holds:

$$\liminf_{\tau, s_k \to 0^+} \frac{\partial}{\partial s} \tilde{\mathscr{L}}(\tau, \tilde{u}_{t+s_k}, \lambda; \Omega_t) \geq \frac{\partial}{\partial s} \tilde{\mathscr{L}}(0, \tilde{u}_t, \lambda; \Omega_t) \quad \forall \lambda \in \tilde{K}_t^\star; \tag{42}$$

(A4) as $s \to 0^+$ an accumulation point $\tilde{\lambda}_t \in \tilde{K}_t^\star$ and a sequence $\tilde{\lambda}_{t+s_k} \in \tilde{K}_{t+s_k}^\star$ exist such that

$$\tilde{\lambda}_{t+s_k} \to \tilde{\lambda}_t \quad \text{strongly in } \mathfrak{T}_{H^\star} - \text{topology} \tag{43}$$

and the upper estimate holds:

$$\limsup_{\tau, s_k \to 0^+} \frac{\partial}{\partial s} \tilde{\mathscr{L}}(\tau, u, \tilde{\lambda}_{t+s_k}; \Omega_t) \leq \frac{\partial}{\partial s} \tilde{\mathscr{L}}(0, u, \tilde{\lambda}_t; \Omega_t) \quad \forall u \in \tilde{K}_t. \tag{44}$$

Then Theorem 2 follows the main theorem on shape differentiability of Lagrangians.

Theorem 3 *Under assumptions (A1)–(A4) there exists a saddle-point* $(\tilde{u}_t, \tilde{\lambda}_t)$ *of* $\frac{\partial}{\partial s} \tilde{\mathscr{L}}(0, u, \lambda; \Omega_t)$ *on* $\tilde{K}_t \times \tilde{K}_t^\star$ *such that*

$$\sup_{\lambda \in \tilde{K}_t^\star} \inf_{u \in \tilde{K}_t} \frac{\partial}{\partial s} \tilde{\mathscr{L}}(0, u, \lambda; \Omega_t) = \frac{\partial}{\partial s} \tilde{\mathscr{L}}(0, \tilde{u}_t, \tilde{\lambda}_t; \Omega_t) = \inf_{u \in \tilde{K}_t} \sup_{\lambda \in \tilde{K}_t^\star} \frac{\partial}{\partial s} \tilde{\mathscr{L}}(0, u, \lambda; \Omega_t) \tag{45}$$

and the shape derivative (39) *is represented by the partial derivative:*

$$\partial_s \tilde{\ell}(0; t) = \frac{\partial}{\partial s} \tilde{\mathscr{L}}(0, \tilde{u}_t, \tilde{\lambda}_t; \Omega_t) \quad \textit{(one-sided)}. \tag{46}$$

Under the assumption (A0) the shape derivatives defined in (25) *and* (39) *coincide:*

$$\partial_t \mathscr{L}(u_t, \lambda_t; \Omega_t) = \partial_s \tilde{\ell}(0; t). \tag{47}$$

4 Application to State-Constrained Shape Optimization

As application of Theorem 3 we consider a least-square minimization subject to state-constraint which approximates overdetermined boundary value problems.

Let a reference domain Ω_t be contained in the hold-all domain $D \subset \mathbb{R}^d$, the velocity $\Lambda(t, x) \in C([t_0, t_1]; W^{1,\infty}(D; \mathbb{R}^d))$ be zero at ∂D. We assume that the boundary $\partial \Omega_t$ is Lipschitz-continuous with the outward unit normal vector n^t and consists of disjoint parts $\Gamma_t^{\mathrm{D}}, \Gamma_t^{\mathrm{N}}, \Gamma_t^{\mathrm{O}}$. For given $g(x), z(x) \in H^2(D; \mathbb{R})$ we start with the *overdetermined problem*: find a domain $\Omega_t \subset D$ and a function $u_t(x) \in H^1(\Omega_t; \mathbb{R})$ satisfying

$$\text{the Laplace equation:} \quad -\Delta u_t = 0 \quad \text{in } \Omega_t, \tag{48}$$

$$\text{the Dirichlet condition:} \quad u_t = 0 \quad \text{on } \Gamma_t^{\mathrm{D}}, \tag{49}$$

$$\text{the Neumann condition:} \quad \frac{\partial}{\partial n^t} u_t = g \quad \text{on } \Gamma_t^{\mathrm{N}}, \tag{50}$$

$$\text{overdetermined conditions:} \quad \frac{\partial}{\partial n^t} u_t = g, \quad u_t = z \quad \text{on } \Gamma_t^{\mathrm{O}}. \tag{51}$$

In particular, prescribing both data g, z and varying Γ_t^{O} describes the *Bernoulli free-boundary problem*. The *inverse identification problem* corresponds to prescribed data g and observed z at fixed Γ_t^{O}. In general, all parts Γ_t^{D}, Γ_t^{N}, Γ_t^{O} can be varied.

In the function space taking into account the Dirichlet condition (49):

$$V(\Omega_t) = \{w \in H^1(\Omega_t; \mathbb{R})| \quad w = 0 \text{ a.e. } \Gamma_t^{\mathrm{D}}\}$$

let the *objective* function $D \times V(\Omega_t) \mapsto \mathbb{R}$, $(\Omega_t, w) \mapsto \mathscr{J}$ approximate the Dirichlet condition in (51) by the least-square misfit:

$$\mathscr{J}(w; \Omega_t) := \tfrac{1}{2} \int_{\Gamma_t^{\mathrm{O}}} (w - z)^2 \, dx. \tag{52}$$

We consider the *state variational problem* describing relations (48)–(50) and the Neumann condition in (51): for fixed Ω_t there exists unique $u_t \in V(\Omega_t)$ minimizing the *energy* function:

$$\mathscr{E}(u_t; \Omega_t) = \min_{w \in V(\Omega_t)} \mathscr{E}(w; \Omega_t) := \tfrac{1}{2} \int_{\Omega_t} |\nabla w|^2 \, dx - \int_{\Gamma_t^{\mathrm{N}} \cup \Gamma_t^{\mathrm{O}}} g w \, dS_x, \tag{53}$$

or satisfying the equivalent first-order *optimality condition*:

$$\langle \mathscr{E}_u(u_t; \Omega_t), p \rangle := \int_{\Omega_t} \nabla u_t \cdot \nabla p \, dx - \int_{\Gamma_t^{\mathrm{N}} \cup \Gamma_t^{\mathrm{O}}} g p \, dS_x = 0 \quad \forall p \in V(\Omega_t). \tag{54}$$

The *state-constrained shape optimization problem* implies: find $\Omega_t \subset D$ such that

$$\min_{\Omega_t \in D} \mathscr{J}(w; \Omega_t) \quad \text{subject to } \mathscr{E}_u(w; \Omega_t) = 0. \tag{55}$$

For the *Lagrangian* function $D \times V(\Omega_t) \times V(\Omega_t) \mapsto \mathbb{R}$, $(\Omega_t, w, p) \mapsto \mathscr{L}$ given by

$$\mathscr{L}(w, p; \Omega_t) := \mathscr{J}(w; \Omega_t) - \langle \mathscr{E}_u(w; \Omega_t), p \rangle \tag{56}$$

the corresponding *primal-dual shape optimization* reads: find triple $(\Omega_{t+s}, u_{t+s}, \lambda_{t+s}) \in D \times V(\Omega_{t+s}) \times V(\Omega_{t+s})$ with $s > 0$ such that

$$\mathscr{L}(u_{t+s}, \lambda_{t+s}; \Omega_{t+s}) < \mathscr{L}(u_t, \lambda_t; \Omega_t). \tag{57}$$

The optimality condition for (57) implies that $u_t \in V(\Omega_t)$ solves the primal equation (54), and the dual variable $\lambda_t \in V(\Omega_t)$ solves the *adjoint equation*:

$$\int_{\Omega_t} \nabla w \cdot \nabla \lambda_t \, dx - \int_{\Gamma_t^{\mathrm{O}}} (u_t - z) w \, dS_x = 0 \quad \forall w \in V(\Omega_t) \tag{58}$$

which describes the following boundary value problem:

$$\text{the Laplace equation:} \quad -\Delta \lambda_t = 0 \qquad \text{in } \Omega_t, \tag{59}$$

$$\text{the Dirichlet condition:} \quad \lambda_t = 0 \qquad \text{on } \Gamma_t^{\mathrm{D}}, \tag{60}$$

$$\text{the Neumann condition:} \quad \tfrac{\partial}{\partial n^t} \lambda_t = 0 \qquad \text{on } \Gamma_t^{\mathrm{N}}, \tag{61}$$

$$\text{the Neumann condition:} \quad \tfrac{\partial}{\partial n^t} \lambda_t = u_t - z \qquad \text{on } \Gamma_t^{\mathrm{O}}. \tag{62}$$

The underlying *optimal value function* $\ell : D \mapsto \overline{\mathbb{R}}$,

$$\ell(\Omega_t) := \mathscr{L}(u_t, \lambda_t; \Omega_t) = \mathscr{J}(u_t; \Omega_t) \tag{63}$$

coincides for the Lagrangian $\mathscr{L}$ and for the objective $\mathscr{J}$ since $\mathscr{E}_u(u_t; \Omega_t) = 0$.

We parametrize perturbed domains by the flow $\phi_s : \Omega_t \mapsto \Omega_{t+s}$ according to (26) and choose small $\delta > 0$ such that $\Omega_{t+s} \subset D$ for $s \in [0, \delta)$. By the construction,

(T1) the map $V(\Omega_{t+s}) \mapsto V(\Omega_t)$, $[(v, \mu) \mapsto (v \circ \phi_s, \mu \circ \phi_s)]$ is bijective.

For the *perturbed Lagrangian* $V(\Omega_{t+s}) \times V(\Omega_{t+s}) \mapsto \mathbb{R}$, $(v, \mu) \mapsto \mathscr{L}$,

$$\mathscr{L}(v, \mu; \Omega_{t+s}) := \frac{1}{2} \int_{\Gamma_{t+s}^{\mathrm{O}}} (v - z)^2 \, dS_y - \int_{\Omega_{t+s}} \nabla_y v \cdot \nabla_y \mu \, dy + \int_{\Gamma_{t+s}^{\mathrm{N}} \cup \Gamma_{t+s}^{\mathrm{O}}} g \mu \, dS_y \tag{64}$$

there exists a unique *saddle-point* $(u_{t+s}, \lambda_{t+s}) \in V(\Omega_{t+s}) \times V(\Omega_{t+s})$ such that

$$\mathscr{L}(u_{t+s}, \lambda_{t+s}; \Omega_{t+s}) = \min_{v \in V(\Omega_{t+s})} \max_{\mu \in V(\Omega_{t+s})} \mathscr{L}(v, \mu; \Omega_{t+s}). \tag{65}$$

The *transformed perturbed Lagrangian* $[0, \delta) \times V(\Omega_t) \times V(\Omega_t) \mapsto \mathbb{R}$, $(s, w, p) \mapsto \tilde{\mathscr{L}}$

$$\tilde{\mathscr{L}}(s, w, p; \Omega_t) := \frac{1}{2} \int_{\Gamma_t^{\mathrm{O}}} (w - z \circ \phi_s)^2 \, \omega_s dS_x$$
$$- \int_{\Omega_t} \big((\nabla \phi_s^{-T} \circ \phi_s) \nabla w \cdot (\nabla \phi_s^{-T} \circ \phi_s) \nabla p \big) \, J_s dx + \int_{\Gamma_t^{\mathrm{N}} \cup \Gamma_t^{\mathrm{O}}} (g \circ \phi_s) p \, \omega_s dS_x, \tag{66}$$

where $J_s := \det(\nabla \phi_s)$ and $\omega_s := |(\nabla \phi_s^{-T} \circ \phi_s) n^t| J_s$ obeys the unique saddle-point

$$(\tilde{u}_{t+s}, \tilde{\lambda}_{t+s}) := (u_{t+s} \circ \phi_s, \lambda_{t+s} \circ \phi_s) \in V(\Omega_t) \times V(\Omega_t) \tag{67}$$

solving the transformed minimax problem

$$\tilde{\mathscr{L}}(s, \tilde{u}_{t+s}, \tilde{\lambda}_{t+s}; \Omega_t) = \min_{w \in V(\Omega_t)} \max_{p \in V(\Omega_t)} \tilde{\mathscr{L}}(s, w, p; \Omega_t). \tag{68}$$

Therefore, the transformed perturbed Lagrangian has the following traits:

(T2) due to (67) the solution sets $\tilde{K}_{t+s} = \{\tilde{u}_{t+s}\}$, $\tilde{K}^{\star}_{t+s} = \{\tilde{\lambda}_{t+s}\}$ are singleton, and the set of saddle-points $\tilde{\mathfrak{S}}(t+s)$ defined in (38) is nonempty for all $s \in [0, \delta)$;

(T3) for all $(w, p) \in V(\Omega_t) \times V(\Omega_t)$ there exists the *partial derivative* at $s = 0$:

$$\begin{aligned} \tfrac{\partial}{\partial s}\tilde{\mathscr{L}}(0, w, p; \Omega_t) &= \int_{\Gamma_t^{\mathrm{O}}} \Big(\tfrac{1}{2}(\mathrm{div}_{\Gamma} \Lambda)(w - z)^2 - (\Lambda \cdot \nabla z)(w - z)\Big)\, dS_x \\ &\quad - \int_{\Omega_t} \Big((\mathrm{div} \Lambda)(\nabla w \cdot \nabla p) - \nabla w \cdot ((\nabla \Lambda + \nabla \Lambda^{\top})\nabla p)\Big)\, dx \\ &\quad + \int_{\Gamma_t^{\mathrm{N}} \cup \Gamma_t^{\mathrm{O}}} \Big((\mathrm{div}_{\Gamma} \Lambda) g + \Lambda \cdot \nabla g\Big) p \, dS_x \end{aligned} \tag{69}$$

with the velocity $\Lambda(t, x)$, where $\mathrm{div}_{\Gamma} \Lambda = \mathrm{div} \Lambda - (\nabla \Lambda n^t) \cdot n^t$, and the asymptotic expansion holds:

$$\tilde{\mathscr{L}}(s, w, p; \Omega_t) = \tilde{\mathscr{L}}(0, w, p; \Omega_t) + s \tfrac{\partial}{\partial s}\tilde{\mathscr{L}}(0, w, p; \Omega_t) + \mathrm{o}(s);$$

(T4) as $s \to 0^+$ there exists a subsequence s_k such that

$$(\tilde{u}_{t+s_k}, \tilde{\lambda}_{t+s_k}) \to (u_t, \lambda_t) \quad \text{strongly in } V(\Omega_t) \times V(\Omega_t). \tag{70}$$

Indeed, the representation (69) follows directly from (66) due to the expansions:

$$\begin{aligned} \nabla \phi_s^{-1} \circ \phi_s = I - s \nabla \Lambda + \mathrm{o}(s), \quad J_s = 1 + s \mathrm{div} \Lambda + \mathrm{o}(s), \quad \omega_s = 1 + s \mathrm{div}_{\Gamma} \Lambda + \mathrm{o}(s), \\ z \circ \phi_s = z + s \Lambda \cdot \nabla z + \mathrm{o}(s), \quad g \circ \phi_s = g + s \Lambda \cdot \nabla g + \mathrm{o}(s), \end{aligned} \tag{71}$$

and the proof of (70) can be found, for example, in [8].

Theorem 4 *For the* parametrized optimal value function $[0, \delta) \mapsto \overline{\mathbb{R}}$, $s \mapsto \tilde{\ell}(s; \Omega_t)$ $:= \tilde{\mathscr{L}}(s, \tilde{u}_{s+t}, \tilde{\lambda}_{s+t}; \Omega_t)$ *there exists the shape derivative represented by*

$$\partial_t \ell(\Omega_t) = \partial_s \tilde{\ell}(0; \Omega_t) = \tfrac{\partial}{\partial s}\tilde{\mathscr{L}}(0, u_t, \lambda_t; \Omega_t) \quad \textit{(one-sided)}. \tag{72}$$

Proof Traits (T1)–(T3) yield (A0)–(A2), the partial derivative $\frac{\partial}{\partial s}\tilde{\mathscr{L}}(\tau, u, \lambda; \Omega_t)$ at $s = \tau$ in (A2) is determined by formula (69) with the respective velocity $\Lambda(t + \tau, x)$. The trait (T4) implies the strong convergences (41) and (43), while continuity of the map $(w, p) \mapsto \frac{\partial}{\partial s}\tilde{\mathscr{L}}(0, w, p; \Omega_t)$ in (69) follows (42) and (44) in assumptions (A3) and (A4). □

Assuming a piecewise $C^{1,1}$-boundary $\partial\Omega_t$ without singular points such that the saddle-point $(u_t, \lambda_t) \in H^2(\Omega_t; \mathbb{R}) \times H^2(\Omega_t; \mathbb{R})$ in Theorem 4, the integration by parts over the domain Ω_t due to (48)–(51) and (59)–(62) yields the expression

$$\int_{\Omega_t} \big((\mathrm{div}\Lambda)(\nabla u_t \cdot \nabla\lambda_t) - \nabla u_t \cdot ((\nabla\Lambda + \nabla\Lambda^{\top})\nabla\lambda_t)\big)\, dx$$

$$= \int_{\partial\Omega_t} \Lambda \cdot \big(n^t(\nabla u_t \cdot \nabla\lambda_t) - \nabla u_t \tfrac{\partial\lambda_t}{\partial n^t} - \nabla\lambda_t \tfrac{\partial u_t}{\partial n^t}\big)\, dS_x. \quad (73)$$

At $\partial\Omega_t$ we decompose $\Lambda = \Lambda_\Gamma + n^t(\Lambda \cdot n^t)$ such that $\mathrm{div}_\Gamma\Lambda = \mathrm{div}_\Gamma\Lambda_\Gamma + (\Lambda \cdot n^t)\varkappa_t$ with the curvature $\varkappa_t := \mathrm{div}_\Gamma n^t$. After integration by parts of the tangential term $\mathrm{div}_\Gamma\Lambda_\Gamma$ with the help of formulas from [28, Sect. 4.4] it follows

$$\int_{\Gamma_t^i} (\mathrm{div}_\Gamma\Lambda) u\, dS_x = \int_{\Gamma_t^i} \Lambda \cdot \big(n^t(u\varkappa_t + \tfrac{\partial u}{\partial n^t}) - \nabla u\big)\, dS_x$$

$$+ \int_{\partial\Gamma_t^i} \Lambda \cdot (\tau^t \times n^t) u\, dL_x, \quad i = \mathrm{N}, \mathrm{O}, \quad (74)$$

where τ^t is a tangential vector at $\partial\Gamma_t^i$ positive oriented to n^t. Using (73) and (74), from (69) we derive the *Hadamard structure representation*:

$$\frac{\partial}{\partial s}\tilde{\mathscr{L}}(0, u_t, \lambda_t; \Omega_t) = \int_{\Gamma_t^{\mathrm{D}}} \Lambda \cdot a^t\, dS_x + \int_{\Gamma_t^{\mathrm{O}}} (\Lambda \cdot n^t) b_t\, dS_x + \int_{\Gamma_t^{\mathrm{N}}} (\Lambda \cdot n^t) c_t\, dS_x$$

$$+ \int_{\partial\Gamma_t^{\mathrm{O}}} \Lambda \cdot (\tau^t \times n^t) G_t\, dL_x + \int_{\partial\Gamma_t^{\mathrm{N}}} \Lambda \cdot (\tau^t \times n^t) g\lambda_t\, dL_x \quad (75)$$

over the boundary, with the components

$$a^t := (\nabla u_t \tfrac{\partial\lambda_t}{\partial n^t} + \nabla\lambda_t \tfrac{\partial u_t}{\partial n^t}) - n^t(\nabla u_t \cdot \nabla\lambda_t), \quad b_t := \varkappa_t G_t + \tfrac{\partial G_t}{\partial n^t} - \nabla u_t \cdot \nabla\lambda_t,$$

$$c_t := \varkappa_t g\lambda_t + \tfrac{\partial(g\lambda_t)}{\partial n^t} - \nabla u_t \cdot \nabla\lambda_t, \quad G_t := \tfrac{1}{2}(u_t - z)^2 + g\lambda_t. \quad (76)$$

This suggests the strategy for numerical shape optimization as follows:

- For fixed Ω_t set a velocity $\Lambda(x)$ at the parts of the boundary $\partial\Omega_t$ either $\Lambda = 0$ or

$$\begin{aligned} \Lambda &= -a^t \quad \text{on } \Gamma_t^{\mathrm{D}}, \\ \Lambda \cdot n^t = -b_t \text{ and } \Lambda_\Gamma &= 0 \quad \text{on } \Gamma_t^{\mathrm{O}}, \\ \Lambda \cdot n^t = -c_t \text{ and } \Lambda_\Gamma &= 0 \quad \text{on } \Gamma_t^{\mathrm{N}}, \\ \Lambda \cdot (\tau^t \times n^t) &= -G_t \quad \text{on } \partial\Gamma_t^{\mathrm{O}}, \\ \Lambda \cdot (\tau^t \times n^t) &= -g\lambda_t \quad \text{on } \partial\Gamma_t^{\mathrm{N}} \end{aligned}$$

providing a *descent direction* by the virtue of (76):

$$\begin{aligned} \tfrac{\partial}{\partial s}\tilde{\mathscr{L}}(0, u_t, \lambda_t; \Omega_t) = -\int_{\Gamma_t^{\mathrm{N}}} |a_t|^2\, dS_x - \int_{\Gamma_t^{\mathrm{D}}} b_t^2\, dS_x - \int_{\Gamma_t^{\mathrm{O}}} c_t^2\, dS_x \\ - \int_{\partial\Gamma_t^{\mathrm{O}}} G_t^2\, dL_x - \int_{\partial\Gamma_t^{\mathrm{N}}} (g\lambda_t)^2\, dL_x < 0; \end{aligned}$$

- find a domain $\Omega_{t+s} \subset D$ bounded by

$$\partial\Omega_{t+s} = \{y \in \mathbb{R}^d |\ \ y = x + s\Lambda(x),\ x \in \partial\Omega_t\} \tag{77}$$

  with a suitable parameter $s > 0$ minimizing the objective $\mathscr{J}(u_{t+s}; \Omega_{t+s})$;
- reset $\Omega_t := \Omega_{t+s}$ and iterate.

Acknowledgments V.A.K. is supported by the Austrian Science Fund (FWF) project P26147-N26: "Object identification problems: numerical analysis" (PION) and the Austrian Academy of Sciences (OeAW). K.O. is supported by the JSPS KAKENHI Grant Number 16K05285. The joint work began in CoMFoS18 that is the workshop by the Activity group MACM (Mathematical Aspects of Continuum Mechanics) of JSIAM. The authors thank the Japan Society for the Promotion of Science (JSPS) research project (No. J19-721) joint with the Russian Foundation for Basic Research (RFBR) project (N 19-51-50004)

References

1. Alekseev, G.V., Mashkov, D.V., Yashenko, E.N.: Analysis of some identification problems for the reaction-diffusion-convection equation. IOP Conf. Ser.: Mater. Sci. Eng. **124**, 012037 (2016)
2. Azegami, H., Ohtsuka, K., Kimura, M.: Shape derivative of cost function for singular point: evaluation by the generalized J integral. JSIAM Lett. **6**, 29–32 (2014)
3. Cakoni, F., Kovtunenko, V.A.: Topological optimality condition for the identification of the center of an inhomogeneity. Inverse Probl. **34**, 035009 (2018)
4. Correa, R., Seeger, A.: Directional derivative of a minimax function. Nonlinear Anal. **9**, 13–22 (1985)
5. Delfour, M.C., Zolésio, J.-P.: Shape and Geometries: Metrics, Analysis, Differential Calculus, and Optimization. SIAM, Philadelphia (2011)
6. González Granada, J.R., Gwinner, J., Kovtunenko, V.A. On the shape differentiability of objectives: a Lagrangian approach and the Brinkman problem. Axioms **7**, 76 (2018)

7. Hasanov Hasanoğlu, A., Romanov, V.G.: Introduction to Inverse Problems for Differential Equations. Springer, Berlin (2017)
8. Haslinger, J., Ito, K., Kozubek, T., Kunisch, K., Peichl, G.: On the shape derivative for problems of Bernoulli type. Interfaces Free Bound. **11**, 317–330 (2009)
9. Hintermüller, M., Kovtunenko, V.A.: From shape variation to topology changes in constrained minimization: a velocity method-based concept. Optim. Methods Softw. **26**, 513–532 (2011)
10. Itou, H., Kovtunenko, V.A., Rajagopal, K.R.: Nonlinear elasticity with limiting small strain for cracks subject to non-penetration. Math. Mech. Solids **22**, 1334–1346 (2017)
11. Itou, H., Kovtunenko, V.A., Tani, A.: The interface crack with Coulomb friction between two bonded dissimilar elastic media. Appl. Math. **56**, 69–97 (2011)
12. Khludnev, A.M., Kovtunenko, V.A.: Analysis of Cracks in Solids. WIT-Press, Southampton, Boston (2000)
13. Khludnev, A.M., Kovtunenko, V.A., Tani, A.: Evolution of a crack with kink and non-penetration. J. Math. Soc. Jpn. **60**, 1219–1253 (2008)
14. Khludnev, A.M., Ohtsuka, K., Sokolowski, J.: On derivative of energy functional for elastic bodies with cracks and unilateral conditions. Q. Appl. Math. **60**, 99–109 (2002)
15. Khludnev, A.M., Sokolowski, J.: Modelling and Control in Solid Mechanics. Birkhäuser, Basel (1997)
16. Kovtunenko, V.A.: Primal-dual methods of shape sensitivity analysis for curvilinear cracks with non-penetration. IMA J. Appl. Math. **71**, 635–657 (2006)
17. Kovtunenko, V.A., Kunisch, K.: Problem of crack perturbation based on level sets and velocities. Z. angew. Math. Mech. **87**, 809–830 (2007)
18. Kovtunenko, V.A., Kunisch, K.: High precision identification of an object: Optimality-conditions-based concept of imaging. SIAM J. Control Optim. **52**, 773–796 (2014)
19. Kovtunenko, V.A., Leugering, G.: A shape-topological control problem for nonlinear crack - defect interaction: the anti-plane variational model. SIAM J. Control Optim. **54**, 1329–1351 (2016)
20. Kovtunenko, V.A., Ohtsuka, K.: Shape differentiability of Lagrangians and application to Stokes problem. SIAM J. Control Optim. **56**, 3668–3684 (2018)
21. Lavrentiev, M.M., Avdeev, A.V., Lavrentiev, M.M.-jr., Priimenko, V.I.: Inverse Problems of Mathematical Physics. De Gruyter (2012)
22. Lazarev, N.P., Rudoy, E.M.: Shape sensitivity analysis of Timoshenko's plate with a crack under the nonpenetration condition. Z. angew. Math. Mech. **94**, 730–739 (2014)
23. Maharani, A.U., Kimura, M., Azegami, H., Ohtsuka, K., Armanda, I.: Shape optimization approach to a free boundary problem. In: V. Suendo, M. Kimura, R. Simanjuntak, S. Miura (eds.) Recent Development in Computational Sciences (Proc. ISCS 2015), vol. 6, pp. 42–55, Kanazawa e-Publishing (2015)
24. Ohtsuka, K.: Shape optimization by generalized J-integral in Poisson's equation with a mixed boundary condition. In: P. van Meurs, M. Kimura, H. Notsu (eds.) Mathematical Analysis of Continuum Mechanics and Industrial Applications II (Proc. CoMFoS16), pp. 73–83. Springer, Berlin (2018)
25. Ohtsuka, K., Kimura, M.: Differentiability of potential energies with a parameter and shape sensitivity analysis for nonlinear case: the p-poisson problem. Jpn. J. Ind. Appl. Math. **29**, 23–35 (2012)
26. Shcherbakov, V.V.: Shape derivative of the energy functional for the bending of elastic plates with thin defects. J. Phys. Conf. Ser. **894**, 012084 (2017)
27. Sokolowski, J., Zolésio, J.-P.: Introduction to Shape Optimization. Shape Sensitivity Analysis. Springer, Berlin (1992)
28. Walker, S.W.: The Shapes of Things: A Practical Guide to Differential Geometry and the Shape Derivative. SIAM, Philadelphia (2015)

Identification of an Unknown Shear Force in Euler–Bernoulli Beam Based on Boundary Measurement of Rotation

Alemdar Hasanov Hasanoglu

Abstract An inverse problem of identifying an unknown shear force $g(t)$ from the measured rotation $\theta(t) := u_x(l, t)$ at the boundary $x = l$ is studied in a system governed by the general form Euler–Bernoulli beam equation $\rho(x)u_{tt} + \mathrm{d}(x)u_t + (r(x)u_{xx})_{xx} - T_r(x)u_{xx} = 0$, $(x, t) \in (0, l) \times (0, T)$ subject to the boundary conditions $u(0, t) = u_x(0, t) = u_{xx}(l, t) = 0$, $((r(x)u_{xx}(x, t))_x)_{x=0} = g(t)$. The Neumann-to-Dirichlet operator $\Phi[\cdot] : \mathscr{G} \subset H^2(0, l) \mapsto L^2(0, T)$ corresponding to this inverse problem is introduced. Compactness and Lipschitz continuity of this operator is proved. An important implication of the last property is that it allows us to prove an existence of a quasi-solution of the inverse problem. An explicit formula for the Fréchet gradient of the Tikhonov functional $J(g) := \|\Phi g - \theta\|^2_{L^2(0,T)}$ is derived through the unique solution of the corresponding adjoint problem. This enables the use of fast gradient methods for numerical solving of the inverse problem.

1 Introduction

In this paper, we study the *inverse problem of identifying the unknown transverse shear force* $g(t)$ in

$$\begin{cases} \rho(x)u_{tt} + \mu(x)u_t + (r(x)u_{xx})_{xx} - T_r u_{xx} = 0, \ (x, t) \in \Omega_T, \\ u(x, 0) = 0, \ u_t(x, 0) = 0, \ x \in (0, l), \\ u(0, t) = u_x(0, t) = 0, \\ (u_{xx}(x, t))_{x=l} = 0, \ (-(r(l)u_{xx}(l, t))_x)_{x=l} = g(t), \ t \in (0, T), \end{cases} \tag{1}$$

from the measured rotation (slope) $\theta(t)$ at the right boundary $x = l$ of a beam:

A. H. Hasanoglu (✉)
Department of Mathematics, Kocaeli University, Izmit, 41001 Kocaeli, Turkey
e-mail: alemdar.hasanoglu@gmail.com

JSPS Fellow, Department of Mathematics, Tokyo University of Science, 1-3 Kagurazaka, Shinjuku-ku, Tokyo 162-8601, Japan

H. Itou et al. (eds.), *Mathematical Analysis of Continuum Mechanics and Industrial Applications III*, Mathematics for Industry 34,
https://doi.org/10.1007/978-981-15-6062-0_8

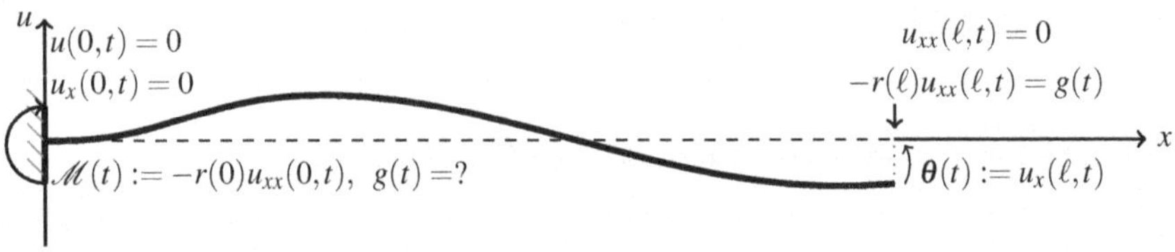

Fig. 1 Geometry of the forward and inverse problems

$$\theta(t) := u_x(l,t), \quad t \in [0,T]. \tag{2}$$

Here $\Omega_T := (0,l) \times (0,T)$, $r(x) = EI(x)$, $E > 0$ is the elasticity modulus, $I(x) > 0$ is the moment of inertia of the cross section, $\rho(x) > 0$ is the mass density of the beam, $\mu(x) > 0$ is the damping coefficient, and $T_r > 0$ is the traction force along the cantilever beam (Fig. 1).

The cantilever beam is an important simplified model for many engineering problems in the fields of mechanical and civil engineering. More recently, cantilevers are found in medical diagnostics and nanoscale measurement systems including a Transverse Dynamic Force Microscope (TDFM). The mathematical model of the problem of estimating the shear force affecting the tip of the cantilever in TDFM with a real-time implementable sliding mode observer has been proposed in [1, 2, 12]:

$$\begin{cases} (EI(Y + \alpha Y_t))_{xxxx} + \rho A_s \dot{Y}_{tt} + \gamma\,\omega \dot{Y}_t = 0,\ x \in (0,L),\ t > 0, \\ Y(0,t) = u(t),\ Y_x(0,t) = 0,\ \ Y_{xx}(l,t) = 0,\ \ -(EIY_{xxx}(L,t)) = f(t),\ t > 0, \end{cases} \tag{3}$$

where $\alpha > 0$ is the internal damping constant of the cantilever. Based on this model and the measured output $y(t) = Y(t,L)$, here the unknown shear force $f(t)$ is estimated. This estimate allows better interpretation and understanding of the scan result. An inverse random source problem related to Euler–Bernoulli beam model has also been studied in [3] to describe the elastic deformation of nanobelts obtained from an atomic force microscope (AFM) under contact model. The model used here is based on the simplest Euler–Bernoulli equation $(E(x)\,I(x)w_{xx})_{xx} = F(x)$, with the load force containing the randomness. Based upon a derived analytical solution for the forward problem, a reconstruction formulae for the inverse problem, i.e., quantifying the constant elastic modulus $E > 0$ and the structure of the random source, is derived.

The problem of identifying the unknown variable coefficients $\rho(x) > 0$ and $r(x) > 0$ in the simplest dynamic Euler–Bernoulli equation $\rho(x)u_{tt} + (r(x)u_{xx})_{xx} = 0$ from the boundary input $g(t)$ and the available observations $u(1,t)$ and $u_x(1,t)$ has been studied in [4]. Assuming that the coefficients are smooth enough, i.e., $\rho, r \in C^4(0,l)$ and the time interval is infinite, i.e., $t \in (0,\infty)$, it is proved that the unknown coefficients $\rho(x) > 0$ and $r(x) > 0$ can be determined uniquely by the triple $\langle g(t), u(1,t), u_x(1,t)\rangle$. In this model, the identification is possible in the case, when an observation was collected in an unbounded time interval $(0, \infty)$ since the method used here is based on the Laplace transform in time of the impulse response. However, in real engineering models, the time interval is finite and may be small enough, since the speed of information propagation in elastic beams is not finite [11]. Thus, the problem of identifying an unknown shear force in a vibrating Euler–Bernoulli beam from available boundary measurements (measured deflection or slope) given in a finite time interval has not been studied in the literature, as far as we know.

The proposed in this paper model (1) and (2) is important not only as a generalization of existing mathematical models in the sense that (a) the Euler–Bernoulli equation (1) includes all the physically possible variable terms; (b) the time interval is finite; and (c) the coefficients in equation (1) may not be smooth enough. The inverse problem (1) and (2) is itself one of the most important problems in vibration theory and applications from the viewpoint of the Neumann-to-Dirichlet map. Indeed, for each admissible input $g(t)$, there exists (in appropriate sense) a unique solution $u(x, t; g)$ of the initial boundary value problem (1), functionally depending on $g(t)$. If $u_x(l, t; g) = \theta(t)$, we say that $g(t)$ is a solution of the inverse problem (1) and (2). Since $g(t)$ is the Neumann input in (1) and $u_x(0, t; g)$ in (2) is the Dirichlet output, the operator $(\Phi g)(t) := u_x(l, t; g)$ has a meaning of the Neumann-to-Dirichlet map corresponding to the inverse problem (1) and (2).

Remark that inverse source problems for the general form Euler–Bernoulli equation (1) defined in a finite domain have been studied in [6–8].

The remainder of this paper is divided into six sections. In Sect. 2, necessary estimates for a weak and a regular weak solutions of the direct problem (1) are derived. Compactness of the Neumann-to-Dirichlet operator is proved in Sect. 3. In Sect. 4, the Tikhonov functional corresponding to the inverse problem (1) and (2) is introduced. It is proved that this functional is Fréchet differentiable. In the case when $T_r = 0$, this result allows to derive an explicit formulae for the Fréchet gradient through the adjoint problem solution. The Lipschitz continuity of the Tikhonov functional is derived in Sect. 5. Using this result, an existence of a minimizer of the Tikhonov functional is proved. Some concluding remarks are given in Sect. 6.

2 Necessary Estimates for the Solution of the Direct Problem (1)

We assume that the inputs $\rho(x), r(x), \mu(x)$, and $g(t)$ satisfy the following conditions:

$$\begin{cases} \rho, r, \mu \in L^\infty(0, l), \ g \in H^1(0, T), \ g(0) = 0, \\ 0 < \rho_0 \le \rho(x) \le \rho_1, \ 0 < r_0 \le r(x) \le r_1, \\ 0 < \mu_0 \le \mu(x) \le \mu_1, \ T_r \ge 0, \ \ g(t) > 0. \end{cases} \tag{4}$$

Under these conditions, there exists a unique weak solution of the direct problem (1) defined as $u \in L^2(0, T; \mathscr{V}(0, l)), u_t \in L^2(0, T; L^2(0, l)), u_{tt} \in L^2(0, T; H^{-2}(0, l))$, where $\mathscr{V}(0, l) := \{v \in H^2(0, l) : v(0) = v'(0) = 0\}$ [7, 8]. This weak solution also belongs to $C([0, T]; H^1(0, l))$, as it follows from Theorem 4, Sect. 5.9 of [5].

We define first the *set of admissible inputs* (transverse shear forces) $\{g(t)\}$ satisfying conditions (4) in the Sobolev space $H^1(0, T)$:

$$\mathbf{G} = \{g \in H^1(0, T) : g(0) = 0, \ 0 < g(t) \leq g^* < \infty\}. \tag{5}$$

Evidently, $\mathbf{G}$ is a nonempty closed convex set in $H^1(0, T)$.

Now we require, in addition to the above conditions, that the inputs $r(x)$ and $g(t)$ satisfy also the following regularity and consistency conditions:

$$g \in H^2(0, T), \ g'(0) = 0. \tag{6}$$

Under conditions (4) and (6), there exists a unique regular weak solution of the direct problem (1) defined as $u \in C(0, T; H^4(0, l))$, $u_t \in C(0, T; \mathscr{V}(0, l))$, $u_{tt} \in L^2(0, T; L^2(0, l))$. Based on these conditions, we define now the *set of admissible inputs* $\mathscr{G}$ in the Sobolev space $H^2(0, T)$ as follows:

$$\mathscr{G} = \{g \in H^2(0, T) : g(0) = g'(0) = 0, \ 0 < g(t) \leq g^* < \infty\}. \tag{7}$$

Lemma 1 *Let conditions (4) hold. Then for the weak solution of the direct problem (1), the following estimates hold:*

$$\|u_{xx}\|^2_{L^2(0,T;\, L^2(0,l))} \leq C_2^2 \exp(C_1^2 T) \, \|g'\|^2_{L^2(0,T)}, \tag{8}$$

$$\|u_t\|^2_{L^2(0,T;\, L^2(0,l))} \leq C_3^2 \, \|g'\|^2_{L^2(0,T)}, \tag{9}$$

where $C_1^2 = 1 + 2T_r^2/(r_0\, \mu_0)$, $C_2^2 = 2l^3\, T(1+T)/r_0^2$, $C_3^2 = T\left(r_0^2\, C_1^2 C_2^2 \exp\left(C_1^2 T\right) + 2l^3(1+T)\right)/(2r_0\rho_0)$ *and the constants* $\rho_0, r_0, \mu_0 > 0$, $T_r \geq 0$ *are defined in (4).*

Proof Multiply both sides of Eq. (1) by $u_t(x, t)$, use the identity

$$(r(x)u_{xx})_{xx}u_t \equiv [(r(x)u_{xx})_x u_t - r(x)u_{xx}u_{tx}]_x + \frac{r(x)}{2}\left(u_{xx}^2\right)_t, \tag{10}$$

integrate over Ω_t, and use the initial and boundary conditions in (1). Then, we obtain the following integral identity:

$$\int_0^l \left[\rho(x)u_t^2 + r(x)u_{xx}^2\right] \mathrm{d}x + 2\int_0^t\int_0^l \mu(x)u_\tau^2 \mathrm{d}x\mathrm{d}\tau$$

$$= 2T_r \int_0^t\int_0^l u_{xx}u_\tau \mathrm{d}x\mathrm{d}\tau + 2\int_0^t g(\tau)u_\tau(l,\tau)\mathrm{d}\tau, \tag{11}$$

for a.e. $t \in [0, T]$. To transform the second right-hand-side integral, we use the integration by parts formula and then the ε-inequality $2ab \leq (1/\varepsilon)\, a^2 + \varepsilon\, b^2$. We have

$$2\int_0^t g(\tau)u_\tau(l,\tau)\mathrm{d}\tau \leq \frac{1}{\varepsilon_1} g^2(t) + \frac{1}{\varepsilon_1}\int_0^t \left(g'(\tau)\right)^2 d\tau + \varepsilon_1 u^2(l,t) + \varepsilon_1 \int_0^t u^2(l,\tau)\mathrm{d}\tau,$$

$t \in [0, T]$. To estimate the third and fourth right-hand-side terms, here we use the inequality

$$u^2(l,t) \leq \frac{l^3}{2}\int_0^l u_{xx}^2(x,t)\mathrm{d}x, \quad \text{a.e, } t \in [0,T], \tag{12}$$

which can be easily derived for a function $u \in L^2(0, T; \mathscr{V}(0, l))$. Then we conclude

$$2\int_0^t g(\tau)u_\tau(l,\tau)\mathrm{d}\tau \leq \frac{1}{\varepsilon_1} g^2(t) + \frac{1}{\varepsilon_1}\int_0^t \left(g'(\tau)\right)^2 \mathrm{d}\tau$$

$$+\frac{l^3\,\varepsilon_1}{2}\int_0^l u_{xx}^2(x,t)\mathrm{d}x + \frac{l^3\,\varepsilon_1}{2}\int_0^t\int_0^l u_{xx}^2(x,\tau)\mathrm{d}x\mathrm{d}\tau. \tag{13}$$

The first right-hand-side integral in (11) can be estimated in the same way:

$$2T_r \int_0^t\int_0^l u_{xx}u_\tau \mathrm{d}x\mathrm{d}\tau \leq \frac{T_r^2}{\varepsilon_2}\int_0^t\int_0^l u_{xx}^2 \mathrm{d}x\mathrm{d}\tau + \varepsilon_2 \int_0^t\int_0^l u_\tau^2 \mathrm{d}x\mathrm{d}\tau.$$

Using this estimate with (13) in the integral identity (11), after elementary transformations we deduce that

$$\rho_0 \int_0^l u_t^2(x,t)\mathrm{d}x + \left(r_0 - \frac{1}{2} l^3 \varepsilon_1\right) \int_0^l u_{xx}^2(x,t)\mathrm{d}x + (2\mu_0 - \varepsilon_2) \int_0^t \int_0^l u_\tau^2(x,\tau)\mathrm{d}x\mathrm{d}\tau$$

$$\leq \left(\frac{1}{2} l^3 \varepsilon_1 + \frac{T_r^2}{\varepsilon_2}\right) \int_0^t \int_0^l u_{xx}^2 \mathrm{d}x\mathrm{d}\tau + \frac{1}{\varepsilon_1} g^2(t) + \frac{1}{\varepsilon_1} \int_0^t \left(g'(\tau)\right)^2 \mathrm{d}\tau,\ t \in [0,T], \quad (14)$$

for a.e. $t \in [0, T]$. We choose the arbitrary parameters $\varepsilon_1,\ \varepsilon_2 > 0$ from the following conditions: $r_0 - l^3 \varepsilon_1/2 > 0$, $2\mu_0 - \varepsilon_2 > 0$ as follows: $\varepsilon_1 = r_0/l^3$, $\varepsilon_2 = \mu_0$, $T_r > 0$. Using this and the estimate

$$g^2(t) + \int_0^t \left(g'(\tau)\right)^2 \mathrm{d}\tau \leq (1+t) \int_0^t \left(g'(\tau)\right)^2 \mathrm{d}\tau, \quad t \in [0, T]$$

in (14), we finally obtain the main integral inequality:

$$\rho_0 \int_0^l u_t^2(x,t)\mathrm{d}x + \frac{1}{2} r_0 \int_0^l u_{xx}^2(x,t)\mathrm{d}x + \mu_0 \int_0^t \int_0^l u_\tau^2(x,\tau)\mathrm{d}x\mathrm{d}\tau$$

$$\leq \left(\frac{r_0}{2} + \frac{T_r^2}{\mu_0}\right) \int_0^t \int_0^l u_{xx}^2 \mathrm{d}x\mathrm{d}\tau + \frac{l^3(1+t)}{r_0} \int_0^t \left(g'(\tau)\right)^2 \mathrm{d}\tau, \text{ a.e. } t \in [0, T]. \quad (15)$$

The first consequence of (15) is the inequality

$$\int_0^l u_{xx}^2(x,t)\mathrm{d}x \leq C_1^2 \int_0^t \int_0^l u_{xx}^2(x,\tau)\mathrm{d}x\mathrm{d}\tau + \frac{2l^3(1+T)}{r_0^2} \int_0^T \left(g'(t)\right)^2 \mathrm{d}t,$$

for a.e. $t \in [0, T]$, where $C_1 > 0$ is the constant defined in the lemma. Applying the Gronwall–Bellman inequality, we arrive at the first required estimate (8).

The second consequence of (15) is the inequality

$$\rho_0 \int_0^t \int_0^l u_\tau^2 \mathrm{d}x\mathrm{d}\tau \leq \left(\frac{r_0}{2} + \frac{T_r^2}{\mu_0}\right) \int_0^t \int_0^l u_{xx}^2 \mathrm{d}x\mathrm{d}\tau + \frac{l^3(1+T)}{r_0} \int_0^T \left(g'(t)\right)^2 \mathrm{d}t,$$

for a.e. $t \in [0, T]$. Using here estimate (8) for the first right-hand-side integral after elementary transformations, we obtain the second estimate (9).

Corollary 1 *Let the conditions of Lemma 1 hold. Then for the following trace estimates hold:*

$$\|u(l,\cdot)\|^2_{L^2(0,T)} \le \frac{l^3}{2}\, C_2^2 \,\exp(C_1^2\, T)\, \|g'\|^2_{L^2(0,T)}, \tag{16}$$

$$\|u_x(l,\cdot)\|^2_{L^2(0,T)} \le l\, C_2^2 \,\exp(C_1^2\, T)\, \|g'\|^2_{L^2(0,T)}, \tag{17}$$

with the constants $C_1, C_2 > 0$ *defined in Lemma 1.*

Proof Proof follows from the trace inequalities

$$\int_0^T u^2(l,t)\mathrm{d}t \le \frac{l^3}{2} \int_0^T\int_0^l u_{xx}^2(x,t)\mathrm{d}x\,\mathrm{d}t, \quad \int_0^T u_x^2(l,t)\mathrm{d}t \le l \int_0^T\int_0^l u_{xx}^2(x,t)\mathrm{d}x\,\mathrm{d}t$$

and estimate (8).

Now assume that in addition to (4), the input $g(t)$ satisfies also the conditions (6). The following lemma shows that the similar estimates can be derived also for the regular weak solution of the direct problem (1).

Lemma 2 *Let conditions (4) and (6) hold. Then for the regular weak solution of the direct problem (1), the following estimates hold:*

$$\|u_{xxt}\|^2_{L^2(0,T;\,L^2(0,l))} \le C_2^2 \,\exp(C_1^2\, T)\, \|g''\|^2_{L^2(0,T)}, \tag{18}$$

$$\|u_{tt}\|^2_{L^2(0,T;\,L^2(0,l))} \le C_3^2\, \|g''\|^2_{L^2(0,T)}, \tag{19}$$

where $C_1,\ C_2,\ C_3 > 0$ *are the constants defined in Lemma 1.*

Proof Differentiate equation (1) with respect to the time variable $t \in (0,T)$, multiply both sides by $u_{tt}(x,t)$, and use the differential identity (10) with $w(x,t)$ replaced by $u_t(x,t)$. Integrating then over Ω_t and using the initial and boundary conditions in (1), we obtain the following integral identity:

$$\int_0^l \left[\rho(x)u_{tt}^2 + r(x)u_{xxt}^2\right]\mathrm{d}x + 2\int_0^t\int_0^l \mu(x)u_{\tau\tau}^2\mathrm{d}x\mathrm{d}\tau$$

$$= 2T_r \int_0^t\int_0^l u_{xx\tau}u_{\tau\tau}\mathrm{d}x\mathrm{d}\tau + 2\int_0^t g'(\tau)u_{\tau\tau}(l,\tau)\mathrm{d}\tau, \tag{20}$$

for a.e. $t \in [0,T]$. This is exactly the identity (11) with $u(x,t)$ replaced by $u_t(x,t)$. The remaining part of the proof is the same as in Lemma 1.

Corollary 2 *Let the conditions of Lemma 2 hold. Then for the following trace estimate hold:*

$$\|u_{xt}(l,\cdot)\|^2_{L^2(0,T)} \le l\, C_2^2 \,\exp(C_1^2\, T)\, \|g''\|^2_{L^2(0,T)}. \tag{21}$$

Remark 1 In fact, we can prove in the same way slightly better estimates in which the norm $\|\cdot\|^2_{L^2(0,T;\,L^2(0,l))}$ in (8), (9), (18) and (19) is replaced by the norm $\|u_{xx}\|^2_{L^\infty(0,T;\,L^2(0,l))}$.

3 Compactness of the Input–Output Operator

Based on the additional condition (2), we define *the input–output map*, i.e., the *Neumann-to-Dirichlet operator* corresponding to the inverse problem (1) and (2) as follows:

$$\Phi g(t) := u_x(x, t; g)|_{x=l}, \tag{22}$$

where $u = u(x, t; g)$ is the solution of the direct problem (1). Depending on regularity of the weak solution of the direct problem, the input–output map (22) can be defined as an operator acting either from $\mathbf{G} \subset H^1(0, T)$ to $L^2(0, T)$ or from $\mathcal{G} \subset H^2(0, l)$ to $L^2(0, T)$, where the sets $\mathbf{G}$ and $\mathcal{G}$ are defined by (5) and (7), correspondingly. Evidently, in the first case the input–output operator defined on a larger set since $\mathcal{G} \subset \mathbf{G} \subset H^1(0, l)$. In both cases, we can reformulate the inverse problem as the linear operator equation

$$\Phi g(t) = \theta(t), \quad t \in (0, T], \tag{23}$$

defined on $\mathbf{G} \subset H^1(0, l)$ or $\mathcal{G} \subset H^2(0, l)$, where $\theta(t)$ is the noise-free measured output. Each case will be examined separately.

Similar to Lemma 3 of [10], we can prove the following compactness result.

Lemma 3 *Let conditions (4) and (6) hold. Then the Neumann-to-Dirichlet operator* $\Phi[\cdot] : \mathcal{G} \subset H^2(0, l) \mapsto L^2(0, T)$ *defined by (22) is a linear compact operator.*

Proof Let $\{g^{(m)}\} \subset \mathcal{G} \subset H^2(0, T)$, $m = \overline{1, \infty}$, be a bounded sequence of inputs. Denote by $\{u^{(m)}(x, t)\}$, $u^{(m)} := u(x, t; g^{(m)})$, the sequence of corresponding regular weak solutions of the direct problem (1). Then $\{u_x(l, t; g^{(m)})\}$ is the sequence of outputs. We need to prove that this sequence is a relatively compact subset of $L^2(0, T)$. By Rellich theorem, it suffices to show that the sequence $\{u_x^{(m)}(l, t)\}$ is bounded in the norm of the Sobolev space $H^1(0, T)$.

It follows from estimate (21) that

$$\|u_{xt}^{(m)}(l, \cdot)\|^2_{L^2(0,T)} \le l\, C_2^2 \exp(C_1^2 T)\, \|(g^{(m)})''\|^2_{L^2(0,T)}.$$

This means the boundedness of the sequence $\{u_{xt}^{(m)}(l, t)\}$ in the norm of $L^2(0, T)$. By the condition $u(l, 0; g_m) = 0$, the norms $\|u_{xt}^{(m)}(l, \cdot)\|_{L^2(0,T)}$ and $\|u_x^{(m)}(l, \cdot)\|_{H^1(0,T)}$ are equivalent. This implies the boundedness of the sequence of outputs $\{u_x(l, t; g^{(m)})\}$ in the norm of $H^1(0, T)$. Then, it follows from the Sobolev embedding theorem that this sequence is relatively compact in $L^2(0, T)$.

With Corollary 1.3.1 of [9], this lemma implies that *the inverse problem* (1) and (2) *is ill-posed.*

4 Fréchet differentiability of the Tikhonov functional

We introduce the Tikhonov functional

$$J(g) := \frac{1}{2}\|\Phi g - \theta\|^2_{L^2(0,T)},\ g \in \mathcal{G} \tag{24}$$

where Φ is the Neumann-to-Dirichlet operator defined by (22). Then

$$\delta J(g) = \int_0^T [u_x(l,t;g) - \theta(t)]\,\delta u_x(l,t;\delta g)\mathrm{d}t + \frac{1}{2}\int_0^T (\delta u_x(l,t;\delta g))^2\,\mathrm{d}t, \tag{25}$$

is the increment, $\delta u(x,t;\delta g) := u(x,t;g+\delta g) - u(x,t;g)$ and $u_x(l,t;g+\delta g)$, $u_x(l,t;g)$ are the outputs corresponding to the inputs $g,\ \delta g \in \mathcal{G}$.

Lemma 4 *Let conditions (4) and (6) hold. Assume that $g, \delta g \in \mathcal{G}$ are arbitrary inputs. Then the following integral relationship holds:*

$$\begin{aligned} -\int_0^T p(t)\,\delta u_x(l,t)\mathrm{d}t &= \int_0^T \phi(l,t)\,\delta g(t)\,\mathrm{d}t \\ &\quad + T_r\int_0^T [\delta u_x(l,t)\phi(l,t) - \delta u(l,t)\phi_x(l,t)]\,\mathrm{d}t, \end{aligned} \tag{26}$$

where $\phi(x,t)$ is the solution of the adjoint problem

$$\begin{cases} \rho(x)\phi_{tt} - \mu(x)\phi_t + (r(x)\phi_{xx})_{xx} - T_r\phi_{xx} = 0,\ (x,t) \in \Omega_T, \\ \phi(x,T) = 0,\ \phi_t(x,T) = 0,\quad x \in (0,l), \\ \phi(0,t) = 0,\ \phi_x(0,t) = 0, \\ (-r(x)\phi_{xx}(x,t))_{x=l} = p(t),\ (-(r(x)\phi_{xx}(x,t))_x)_{x=l} = 0,\ t \in (0,T), \end{cases} \tag{27}$$

with the arbitrary input $p \in L^2(0,T)$ and $\delta u(x,t;\delta g)$ is the solution of the problem:

$$\begin{cases} \rho(x)\delta u_{tt} + \mu(x)\delta u_t + (r(x)\delta u_{xx})_{xx} - T_r\delta u_{xx} = 0,\ (x,t) \in \Omega_T, \\ \delta u(x,0) = 0,\ \delta u_t(x,0) = 0,\quad x \in (0,l), \\ \delta u(0,t) = \delta u_x(0,t) = 0, \\ (\delta u_{xx}(x,t))_{x=l} = 0,\ (-(r(x)\delta u_{xx}(x,t))_x)_{x=l} = \delta g(t),\ t \in (0,T). \end{cases} \tag{28}$$

Proof Multiply both sides of Eq. (28) by $\phi(x,t)$, integrate over Ω_T, and apply the integration by parts formula multiple times. Then, we obtain

$$\int_0^l \int_0^T \left[\rho(x)\phi_{tt} - \mu(x)\phi_t + (r(x)\phi_{xx})_{xx} - T_r\phi_{xx}\right] \delta u \, \mathrm{d}t\mathrm{d}x$$

$$+ \int_0^l [\rho(x)\delta u_t \phi - \rho(x)\delta u \phi_t + \mu(x)\delta u\, \phi]_{t=0}^{t=T} \, \mathrm{d}x$$

$$+ \int_0^T \left[(r(x)\delta u_{xx})_x \, \phi - r(x)\delta u_{xx}\phi_x + r(x)\delta u_x \phi_{xx} - \delta u\, (r(x)\phi_{xx})_x\right]_{x=0}^{x=l} \mathrm{d}t$$

$$+ \int_0^T [-T_r \delta u_x \phi + T_r \delta u \phi_x]_{x=0}^{x=l} \, \mathrm{d}t = 0. \tag{29}$$

Since $\phi(x,t)$ is the solution of the adjoint problem, the term in the bracket under the integral of the first line of (29) is zero. Taking into account the initial/final and the boundary conditions in (27) and (28), we arrive at the relationship (26).

Choose now the arbitrary input $p(t)$ in the adjoint problem (27) as follows: $p(t) = -[u_x(l,t;g) - \theta(t)]$, $t \in [0,T]$. Then from the integral relationship (26) and the increment formula (25), we deduce that

$$\delta J(g) = \int_0^T \phi(l,t;g)\delta g(t)\mathrm{d}t + T_r \int_0^T [\phi_x(l,t;g)\, \delta u(l,t;\delta g)$$

$$-\phi(l,t;g)\, \delta u_x(l,t;\delta g)]\, \mathrm{d}t + \frac{1}{2}\int_0^T [\delta_x u(l,t;\delta g)]^2 \mathrm{d}t. \tag{30}$$

Theorem 1 *Let conditions (4) hold. Then the Tikhonov functional $J(g)$, $g \in \mathbf{G}$, defined by (24), is Fréchet differentiable. Moreover, in the case when $T_r = 0$, for the Fréchet gradient of this functional the following explicit gradient formula holds:*

$$J'(g)(t) = \phi(l,t;g), \ \text{a.e. } t \in (0,T), \tag{31}$$

where $\phi(x,t)$ is the weak solution of the adjoint problem (27) with $T_r = 0$ and with $p(t) = u_x(l,t;g) - \theta(t)$.

Proof Indeed, it follows from the trace estimates (16) and (17) that the norms $\|u(l,\cdot;g)\|_{L^2(0,T)}$ and $\|u_x(l,\cdot;g)\|_{L^2(0,T)}$ are of the order $\mathcal{O}(\|g'\|_{L^2(0,T)})$.

In the case when $T_r = 0$ the second and third right-hand-side integrals in (30) vanish, and last right-hand-side integral is of the order $\mathcal{O}(\|g'\|^2_{L^2(0,T)})$, which imply formula (31).

5 An Existence of a Minimizer of the Tikhonov Functional

Lemma 5 *Let conditions (4) hold. Then the input–output operator* $\Phi[\cdot] : \mathbf{G} \subset H^1(0,l) \mapsto L^2(0,T)$ *defined by (22) and corresponding to the inverse problem (1) and (2) is Lipschitz continuous, that is,*

$$\|\Phi[g_1] - \Phi[g_2]\|_{L^2(0,T)} \leq L_0 \|g_1' - g_2'\|_{L^2(0,l)}, \; \textit{for all } g_1, g_2 \in \mathbf{G}, \tag{32}$$

with the Lipschitz constant $L_0 = l^{3/2} C_2 \exp(C_1^2 T/2)/\sqrt{2} > 0$, *where* $C_1,\, C_2 > 0$ *are the constants defined in Lemma 1.*

Proof Let $u_k(x,t) := u(x,t;g_k)$, $k = 1, 2$, be weak solutions of the direct problem (1) corresponding to the inputs $g_1, g_2 \in \mathbf{G}$. Then $\delta u(x,t) = u_1(x,t) - u_2(x,t)$ solves the problem (28) with the input $\delta g(t) = g_1(t) - g_2(t)$. By the definition of the input–output operator, we have

$$\|\Phi[g_1] - \Phi[g_2]\|_{L^2(0,T)} = \|\delta u_x(l,\cdot)\|_{L^2(0,T)}.$$

Here, we use estimate (17) in Corollary 1. Then, we get

$$\|\Phi[g_1] - \Phi[g_2]\|^2_{L^2(0,T)} \leq l\, C_2^2 \exp(C_1^2\, T)\, \|\delta g'\|^2_{L^2(0,T)}.$$

This implies the required result.

The Lipschitz continuity of the input–output operator Φ allows to prove existence of a solution of the minimization problem for the Tikhonov functional (24):

$$J(g) = \inf_{\tilde{g} \in \mathbf{G}} J(\tilde{g}). \tag{33}$$

Theorem 2 *Let conditions (4) hold. Then the minimization problem (33) has a solution in* $\mathbf{G} \subset H^1(0,T)$.

Proof The proof follows from the generalized Weierstrass existence theorem (Sect. 2.5, Theorem 2.D, [13]). Indeed, set of admissible inputs $\mathbf{G}$ defined by (5) is a nonempty closed convex set of the Sobolev space $H^1(0,T)$. By Lemma 5, the Tikhonov functional $J(g)$ is Lipschitz continuous, and hence, is lower semicontinuous. As a lower semicontinuous functional defined on the nonempty closed convex set $\mathbf{G}$, it is weakly sequentially lower semicontinuous by Lemma 5, Sect. 2.5 of [13]. Then by the generalized Weierstrass existence theorem, it has a minimum on $\mathbf{G}$.

6 Conclusion

We analyzed the inverse boundary value problem of identifying an unknown shear force $g(t)$ in a system governed by the general form Euler–Bernoulli beam equation, from the measured rotation $\theta(t) := u_x(l, t)$ at the boundary $x = l$. This is a new problem in inverse problems theory and has very important engineering applications. We introduced the Neumann-to-Dirichlet operator corresponding to the inverse problem and derived it from the main properties. These properties allow to not only prove an existence of a quasi-solution of the inverse problem, but also obtain the Fréchet differentiability of the Tikhonov functional. As a consequence, the proposed approach allows also the use of gradient-type iteration algorithms for numerical solving of the inverse problem.

Acknowledgments This research has been supported by the Japan Society of the Promotion of Science through the International Program FY2018 JSPS Individual Fellowship for Research in Japan. Particular thanks are due to Dr. Hiromichi Itou who has provided thoughtful care during my visit to the Department of Mathematics Tokyo University of Science. The author would like to thank the anonymous reviewer for helpful and constructive comments that greatly contributed to improving the final version of the manuscript.

References

1. Antognozzi, M.: Investigation of the shear force contrast mechanism in transverse dynamic force microscopy. Ph.D. Thesis, University of Bristol, UK (2000)
2. Antognozzi, M., Binger, D., Humphris, A., James, P., Miles, M.: Modeling of cylindrically tapered cantilevers for transverse dynamic force microscopy (TDFM). Ultramicroscopy **86**, 223–232 (2001)
3. Bao, G., Xu, X.: An inverse random source problem in quantifying the elastic modulus of nanomaterials. Inverse Prob. **29**, 015006 (16p) (2013)
4. Chang, J.-D., Guo, B.Z.: Identification of variable spatial coefficients for a beam equation from boundary measurements. Automatica **43**, 732–737 (2007)
5. Evans, L.C.: Partial Differential Equations, Rhode Island: American Mathematical Society (2002)
6. Hasanov, A.: Identification of unknown source term in a vibrating cantilevered beam from final overdeterminations. Inverse Prob. **25** 115015 (19p) (2009)
7. Hasanov, A., Kawano, A.: Identification of unknown spatial load distributions in a vibrating Euler-Bernoulli beam from limited measured data. Inverse Prob. **32**(5) 055004 (31p) (2016)
8. Hasanov, A., Baysal, O.: Identification of unknown temporal and spatial load distributions in a vibrating Euler-Bernoulli beam from Dirichlet boundary measured data. Automatica **71**, 106–117 (2016)
9. Hasanov Hasanoglu, A., Romanov, V.G.: Introduction to Inverse Problems for Differential Equations. Springer, New York (2017)
10. Hasanov, A., Baysal, O., Sebu, C.: Identification of an unknown shear force in the Euler-Bernoulli cantilever beam from measured boundary deflection. Inverse Prob. (Accepted). https://doi.org/10.1088/1361-6420/ab2a34

11. Landau, L.D., Lifshitz, E.M.: Theory of Elasticity, 3rd edn. Butterworth-Heinemann, New York (1986)
12. Nguyen, T., et al.: Estimation of the shear force in transverse dynamic force microscopy using a sliding mode observer. AIP Adv. **5** (097157) (2015)
13. Zeidler, E.: Applied Functional Analysis. Main Principles and Their Applications. Springer, New York (1995)

Shape Optimization of Flow Fields Considering Proper Orthogonal Decomposition

Takashi Nakazawa

Abstract This paper presents a new shape optimization method suppressing time periodic flows driven only by the non-stationary boundary condition at a sufficiently low Reynolds number using Proper Orthogonal Decomposition (POD). For the shape optimization problem, the eigenvalue in POD is defined as a cost function. The main problems are the non-stationary Navier–Stokes problem and the eigenvalue problem of POD. An objective functional is described using Lagrange multiplier method with finite element method. Two-dimensional cavity flow with a disk-shaped isolated body is adopted, where the non-stationary boundary condition is defined on the top boundary and non-slip boundary condition for the boundaries not only of the side and bottom but also of the disk. For numerical demonstrations, the disk boundary is used as the design boundary. Therefore the disk is reshaped by the shape optimization process as the cost function decreases. Numerical results reveal that the cost function (eigenvalues in POD) is decreased. The eigenvalues in the initial and the optimal domains are compared. Results clarify suppression of the amplitude of the time periodic flow, driven only by the non-stationary boundary condition at a sufficiently low Reynolds number.

Keywords Cavity flow · Shape optimization problem · Adjoint method · Proper orthogonal decomposition

1 Introduction

This paper presents a solution of a shape optimization problem for suppressing a time periodic flow driven only by the non-stationary boundary condition at a sufficiently low Reynolds number. Recently, a shape optimization method aimed at direct control of flow field stability was constructed by Nakazawa and Azegami [1] based on linear

T. Nakazawa (✉)
Center for Mathematical Modeling and Data Science, Osaka University, 1-3, Machikaneyama-cho, Toyonaka-shi, Osaka 560-8531, Japan
e-mail: nakazawa@sigmath.es.osaka-u.ac.jp

H. Itou et al. (eds.), *Mathematical Analysis of Continuum Mechanics and Industrial Applications III*, Mathematics for Industry 34,
https://doi.org/10.1007/978-981-15-6062-0_9

stability analysis. However, in [1], the stationary Navier–Stokes problem should be solved to obtain the stationary solution for linear stability analysis. Therefore, in the case in which the non-stationary boundary condition is defined, the shape optimization problem addressed in [1] cannot be used. As described in this paper, the author specifically examines construction of a shape optimization method to suppress a time periodic flow driven only by the non-stationary boundary condition at a sufficiently low Reynolds number, based on Proper Orthogonal Decomposition (POD). The new shape optimization problem can be applied even if the non-stationary boundary condition is considered because the non-stationary Navier–Stokes problem should be solved for POD. The particular history, background, and procedure of the suggested shape optimization problem are described below.

In fluid dynamics, Haslinger and Makinen [2], Mohammadi and Pironneau [3], and Moubachir and Zolesio [4] have been developing shape optimization problems. And more, the stationary Stokes problem is used as the main problem by Shinohara [5]. Additionally, Katamine et al. [6] and Ghosh [7] demonstrated the shape optimization problem of a disk-shaped isolated body and a two-dimensional airfoil in a domain defining the stationary Navier–Stokes equations as the main problem. Iwata et al. [8] solved shape optimization problems related to non-stationary Navier–Stokes problems and designed a disk-shaped isolated body located initially in two-dimensional Poiseuille flow at low Reynolds numbers. In the field of aerodynamics, compressible stationary Euler problems are used as the main problem to design two-dimensional and three-dimensional wings in Li [9] and Schmidt [10].

Although flow stabilization presents an important challenge of which research field of flow control is facing, few reports of the relevant literature describe flow stabilization by shape optimization. Thereby, Nakazawa [11] reported that minimization and maximization problems of the dissipation energy are solved in the two-dimensional cavity flow, where the stationary Navier–Stokes problem is used as the main problem and the dissipation energy is used as the cost function. After shape optimization, linear stability analysis in the initial and the optimal domains is performed. The critical Reynolds numbers are, respectively, decreasing and increasing. Next, for controlling the flow stability more directly, Nakazawa and Azegami [1] reported a pioneering shape optimization method to stabilize the disturbances. The method is based on the linear stability theory. In particular, the real part of the leading eigenvalue is used as the cost function. For obtaining the cost function, the stationary Navier–Stokes problem and the eigenvalue problem of the linear stability analysis are defined as the main problems. However, the methods explained above are not available for the case in which the non-stationary boundary condition is defined because the stationary Navier–Stokes problem should be solved in [1, 11].

To meet the challenge explained above, the author constructs a new shape optimization method using Proper Orthogonal Decomposition (POD), in which the eigenvalue in POD is defined as the cost function. A remarkable feature of this suggested shape optimization problem is that a time periodic flow driven only by the non-stationary boundary condition at a sufficiently low Reynolds number is developed or suppressed efficiently in this paper because the eigenvalue (cost function) shows the L^2 norm of the velocity vector which takes maximum and minimum values. A brief summary of the suggested shape optimization problem is presented below.

The sum of eigenvalues in POD is defined as the cost function. The non-stationary Navier–Stokes problem and the eigenvalue problem in POD are used as the main problems. The main problems are transformed from strong forms to weak forms with trial functions based on a standard manner of Finite Element Method (FEM). The functional is described by Lagrange multiplier method with FEM. Next, its first variation (which is the same as the material derivative) is derived to evaluate sensitivity using adjoint variable method. An initial domain is reshaped iteratively to obtain an optimal domain. Then the H^1 gradient method is used for stable domain deformation.

This paper particularly addresses setting of the suggested shape optimization problem. Therefore, a sufficiently low Reynolds number is adopted to avoid consideration of vortices in the boundary layer, where Reynolds number Re = 100 is selected. Appendix D of this paper shows numerically that a secondary flow is not developed.

For numerical demonstrations, two-dimensional cavity flow with a disk-shaped isolated body is adopted, where the non-stationary boundary condition is defined on the top boundary and non-slip boundary condition for the boundaries not only of the side and bottom, but also of the disk. The disk boundary is used as the design boundary. Therefore the disk is reshaped by the shape optimization process as the cost function decreases. After numerical calculations, the eigenvalues of POD are compared in the initial domain and the optimal domain. The effectiveness of the suggested shape optimization problem is confirmed.

2 Formulation of Problem

2.1 Initial Domain

Let Ω_0 be a fixed bounded Lipschitz domain in $\mathbb{R}^d$ ($d \in \mathbb{N}$). Additionally, let Ω be an open subset of Ω_0, a position vector is denoted as $\boldsymbol{x} \in \mathbb{R}^d$. As described in this paper, the two-dimensional cavity flow with a disk-shaped isolated body Ω is adopted for the initial domain. For $d = 2$, the initial domain is $\Omega \subset \Omega_0 \subset \mathbb{R}^2$,

$$\Omega = \Omega_{\mathrm{M}} \backslash \Omega_{\mathrm{m}}, \tag{1}$$

$$\Omega_{\mathrm{M}} = \{(x, y)\ ;\ 0 \leq x \leq 1,\ 0 \leq y \leq 1\}, \tag{2}$$

$$\Omega_{\mathrm{m}} = \left\{(x, y)\ ;\ (x - 0.5)^2 + (y - 0.5)^2 \leq 0.1\right\}, \tag{3}$$

regarding the boundary as

$$\Gamma_{\mathrm{top}} = \{(x, y)\ ;\ 0 \leq x \leq 1,\ y = 1\}, \tag{4}$$

$$\Gamma_{\mathrm{wall}} = \partial\Omega_{\mathrm{M}} \backslash \Gamma_{\mathrm{top}}. \tag{5}$$

For domain reshaping, the boundary of the disk $\partial\Omega_{\mathrm{m}}$ is used as the design boundary.

2.2 Domain Variation

We consider a domain deformation $\boldsymbol{\phi}$ as $\Omega \to \boldsymbol{\phi}(\Omega)$, where $\boldsymbol{\phi}$ is $\mathbb{R}^d$-valued function. For $|\epsilon| \ll 1$, mapping $\boldsymbol{\phi}$ is represented by $\boldsymbol{\phi} = \boldsymbol{\phi}_0 + \epsilon\boldsymbol{\varphi}$ in $W^{1,\infty}(\Omega, \mathbb{R}^d)$. Then we designate it by the identity map $\boldsymbol{\phi}_0(\Omega) = \Omega$ and the domain variation $\boldsymbol{\varphi}$.

We assume ζ as a scalar-valued function describing a physical state in Ω. For such ζ, we introduce the following energy functional

$$L(\Omega, \zeta) = \int_{\Omega} G\,(\boldsymbol{x}, \zeta)\,\mathrm{d}x, \tag{6}$$

where G represents a real-valued given energy density function. The first variation of the functional is expressed as

$$\dot{L}(\bar{\Omega}, \zeta, \boldsymbol{\phi}) = \int_{\Omega} G'\,(\boldsymbol{x}, \zeta)\,\mathrm{d}x + \int_{\partial\Omega} G\,(\boldsymbol{x}, \zeta)\,\boldsymbol{\nu} \cdot \boldsymbol{\varphi}\mathrm{d}\gamma, \tag{7}$$

where $(\dot{\cdot})$ and $(\cdot)'$ depict the material derivative and the Fréchet derivative with respect to ζ ,and $\boldsymbol{\nu}$ denotes the outward unit normal vector on the boundary. Additional details are presented in [12].

Considering the initial domain Ω and the design boundary $\partial\Omega_{\mathrm{m}}, \boldsymbol{\phi} = \mathbf{0}$ on $\partial\Omega_{\mathrm{M}}$, we have the first variation as

$$\dot{L}(\Omega \cup \partial\Omega_{\mathrm{m}}, \zeta, \boldsymbol{\phi}) = \int_{\Omega} G'\,(\boldsymbol{x}, \zeta)\,\mathrm{d}x + \int_{\partial\Omega_{\mathrm{m}}} G\,(\boldsymbol{x}, \zeta)\,\boldsymbol{\nu} \cdot \boldsymbol{\varphi}\mathrm{d}\gamma. \tag{8}$$

where the first integral of G' in (8) is independent of $\boldsymbol{\phi}$.

For sensitivity analysis, the adjoint variable method is used to derive a main problem and an adjoint problem by setting

$$\int_{\Omega} G'\,(\boldsymbol{x}, \zeta)\,\mathrm{d}x = 0. \tag{9}$$

After solving the main and the adjoint problems, density function $G\,(\boldsymbol{x}, \zeta)$ is evaluated as the sensitivity by substituting the main and adjoint variables into Eq. (8).

In fact, for numerical calculations, it was noticed by Imam [13] that direct application of the sensitivity often results in oscillating shapes. The reason for oscillation is why the sensitivity lacks smoothness. To avoid oscillation, a method using the Laplace operator as a smoother, called the H^1 gradient method, was proposed by Azegami et al. [12]. Particularly, the mapping $\boldsymbol{\phi} = \boldsymbol{\phi}_0 + \epsilon\boldsymbol{\varphi}$ should be in $W^{1,\infty}(\Omega, \mathbb{R}^d)$, but the domain variation $\boldsymbol{\varphi}$ substituted the main and adjoint variables is generally in $L^\infty(\Omega, \mathbb{R}^d)$, and $\boldsymbol{\varphi}$ is lack of C^1. Therefore, by the H^1 gradient method, the domain variation is able to recover C^1 class and the domain shape is removed safely as a result. It has been reported from some earlier studies [1, 6, 8, 11, 14] that the objective functional $L(\Omega, \zeta)$ is minimized safely by the H^1 gradient method.

2.3 Domain Reshaping Method

We obtain the optimal domain numerically using the following iterating scheme, where an integer number $k \in \mathbb{N}$ depicts the iteration step, and where positive values $\epsilon, \beta \in \mathbb{R}$ depict an arbitrary small parameters, $K \in \mathbb{N}$ denotes the final step number when the iteration scheme finishes.

Step 1 Set $k = 0$ and $\Omega^k, \zeta^k, \boldsymbol{\phi}^k$.
Step 2 Define the functional $L(\Omega^k, \zeta^k)$, and the density function $G\left(\boldsymbol{x}, \zeta^k\right)$.
Step 3 Derive the first variation $\dot{L}(\Omega^k \cup \partial\Omega^k_{\mathrm{m}}, \zeta^k, \boldsymbol{\phi}^k)$.
Step 4 Evaluate sensitivity $G(\boldsymbol{x}, \zeta^k)\boldsymbol{\nu}$.
Step 5 Operate H^1 gradient method to obtain the smoothed sensitivity $\boldsymbol{\varphi}^k$.
Step 6 Obtain the new domain $\Omega^{k+1} = \boldsymbol{\phi}^{k+1}(\Omega^{k+1}) = \boldsymbol{\phi}_0(\Omega^k) + \epsilon\boldsymbol{\varphi}^k(\Omega^k)$.
Step 7 Judge the convergence:

- If terminal condition $|f^{k+1} - f^k|/f^0 < \beta$ is satisfied for the cost function f, then stop.
- Otherwise, replace $k + 1$ with k and return to Step 2.

In fact, all the variables, functions, function spaces, and functionals used for this study depend on the iteration step k for domain deformation. Hereinafter, for convenience, iteration step k is not described.

2.4 Main Problems

For a shape optimization problem considering POD, the paper is concerned with the main problems, which are the non-stationary Navier–Stokes problem, and the eigenvalue problem in POD. Below, it is assumed that the mapping $\boldsymbol{\phi}$ is given, that an initial domain Ω and the boundaries are determined, and that the flow of a viscous incompressible fluid that occupies a bounded domain Ω in $\mathbb{R}^d$ is studied.

The velocity $\boldsymbol{u}$ and pressure p are assumed to be satisfied in this domain Ω. The Reynolds number Re is defined with reference length $|\Gamma_{\text{top}}|$ and the reference speed which is the maximum value in the x-direction velocity component on Γ_{top}.

2.4.1 The Non-stationary Navier–Stokes Problem

Problem 1 (*Non-stationary Navier–Stokes*) For $\boldsymbol{x} \in \boldsymbol{\phi}(\Omega)$ and time $t \in \mathbb{R}$, find $(\boldsymbol{u}, p) : \Omega \times (0, T) \to \mathbb{R}^d \times \mathbb{R}$ such that

$$\frac{D\boldsymbol{u}}{Dt} = -\nabla p + \frac{1}{\text{Re}} \Delta \boldsymbol{u} \quad \text{in } \Omega \times (0, T), \tag{10}$$

$$\nabla \cdot \boldsymbol{u} = 0 \quad \text{in } \Omega \times (0, T), \tag{11}$$

$$\boldsymbol{u} = \boldsymbol{u}_{\text{D}} \cos(2\pi t) \quad \text{on } \partial\Omega \times (0, T), \tag{12}$$

$$\boldsymbol{u} = \boldsymbol{u}^0 \quad \text{in } \Omega \text{ at } t = 0, \tag{13}$$

where

$$\boldsymbol{u}_{\text{D}} = \boldsymbol{0} \text{ on } \Gamma_{\text{wall}} \cup \partial\Omega_{\text{m}} \text{ and } \boldsymbol{u}_{\text{D}} = \left(16x^2(x-1)^2, 0\right) \text{ on } \Gamma_{\text{top}}, \tag{14}$$

and where $\boldsymbol{u}^0$ represents the stationary solution of the stationary Navier–Stokes problem.

Let $(\boldsymbol{w}, q)$ be the adjoint variables with respect to the velocity and the pressure. By discretizing in the time direction with the finite difference method, a set of necessary variables is written as $\zeta_1 = \{\boldsymbol{u}, p, \boldsymbol{w}, q\}$, where hereinafter $\boldsymbol{u} = \{\boldsymbol{u}^n\}_{n=N_1}^{N_2}$, $\{p^n\}_{n=N_1}^{N_2}$, $\{\boldsymbol{w}^n\}_{n=N_1}^{N_2}$, $\{q^n\}_{n=N_1}^{N_2}$ with integer $N_2 > N_1$. The variational form of the non-stationary Navier–Stokes problem is defined as

$$L_1(\Omega, \zeta_1) = -\frac{1}{m} \sum_{n=N_1}^{N_2} \left\{ \int_{\Omega} G_1^n(\boldsymbol{x}, \zeta_1)\, \mathrm{d}x \right\}, \tag{15}$$

by setting $m = N_2 - N_1 + 1 > 0$ with $N_1 = \frac{T_1}{\Delta t}$ and $N_2 = \frac{T_2}{\Delta t}$ for time step size Δt, at time $t = T_1, T_2$ with suitable $T_2 > T_1$. The density function $G_1^n(\boldsymbol{x}, \zeta_1)$ is presented as

$$G_1^n(\boldsymbol{x}, \zeta_1) = \frac{\boldsymbol{u}^{n+1}(\boldsymbol{x}) - \boldsymbol{u}^n(\boldsymbol{X}^n)}{\Delta t} \cdot \boldsymbol{w}^{n+1}$$
$$- p^{n+1}\nabla \cdot \boldsymbol{w}^{n+1} - q^{n+1}\nabla \cdot \boldsymbol{u}^{n+1} + \frac{1}{\text{Re}}(\nabla \boldsymbol{u}^{n+1})^{\text{T}} : (\nabla \boldsymbol{w}^{n+1})^{\text{T}}, \tag{16}$$

where $\boldsymbol{X}^n = \boldsymbol{x} - \Delta t \boldsymbol{u}^n$, using the characteristic finite element scheme addressed in Notsu [15]. Additional details are available in Appendix A.

2.4.2 Snapshot Proper Orthogonal Decomposition

We define a snapshot Proper Orthogonal Decomposition (Snapshot POD) analysis from time $t = T_1$ to T_2. The correlation coefficient matrix $\mathrm{R} \in \mathbb{R}^{m\times m}$ is formed by

$$\tilde{\boldsymbol{u}} = \left[\boldsymbol{u}^{N_1}, \ldots, \boldsymbol{u}^{N_2}\right] \in \mathbb{R}^{d\times m} \tag{17}$$

as

$$\mathrm{R}(N_1, N_2, \tilde{\boldsymbol{u}}, \tilde{\boldsymbol{u}}) = \int_{\Omega} \tilde{\boldsymbol{u}}^T \tilde{\boldsymbol{u}} \mathrm{d}x.$$

Let eigenvalues and eigenfunctions in POD be $\omega \in \mathbb{R}^m$ and $\hat{\boldsymbol{u}} \in \mathbb{R}^{m\times m}$,

$$\omega = \left[\omega^1, \ldots, \omega^i, \ldots, \omega^m\right],\ \omega^i \in \mathbb{R},$$
$$\hat{\boldsymbol{u}} = \left[\hat{\boldsymbol{u}}^1, \ldots, \hat{\boldsymbol{u}}^i, \ldots, \hat{\boldsymbol{u}}^m\right],\ \hat{\boldsymbol{u}}^i \in \mathbb{R}^m,$$
$$\boldsymbol{\Phi} = \tilde{\boldsymbol{u}}\hat{\boldsymbol{u}}\omega^{-\frac{1}{2}} \in \mathbb{R}^{d\times m},$$

where $\mathrm{R}(N_1, N_2, \tilde{\boldsymbol{u}}, \tilde{\boldsymbol{u}})$ is a positive-semidefinite matrix satisfying the eigenvalue $0 \leq \omega$, and where $\boldsymbol{\Phi}^i$ depicts the POD basis for the ith primary component as

$$\boldsymbol{\Phi} = \left[\boldsymbol{\Phi}^1, \ldots, \boldsymbol{\Phi}^i, \ldots, \boldsymbol{\Phi}^m\right].$$

Using the definitions, we define snapshot POD analysis as described below.

Problem 2 (*Snapshot Proper Orthogonal Decomposition*) Let the solution $\boldsymbol{u}$ of Problem 1 be given and the identify matrix $\boldsymbol{I} \in \mathbb{R}^{m\times m}$. Find $\omega \in \mathbb{R}^m$ and $\hat{\boldsymbol{u}} \in \mathbb{R}^{m\times m}$ for $N_1, N_2 \in \mathbb{N}$ such that

$$\omega \boldsymbol{I} \hat{\boldsymbol{u}} = \mathrm{R}(N_1, N_2, \tilde{\boldsymbol{u}}, \tilde{\boldsymbol{u}})\hat{\boldsymbol{u}}. \tag{18}$$

For the optimization problem, let $\zeta_2 = \left\{\omega, \hat{\boldsymbol{u}}, \boldsymbol{\alpha}, \boldsymbol{u}\right\}$ be the set of necessary variables used in Problem 2, where $\boldsymbol{\alpha}$ is a adjoint variable for the eigenfunction $\hat{\boldsymbol{u}}$.

$$\boldsymbol{\alpha} = \left[\boldsymbol{\alpha}^1, \ldots, \boldsymbol{\alpha}^i, \ldots, \boldsymbol{\alpha}^m\right],\ \boldsymbol{\alpha}^i \in \mathbb{R}^m. \tag{19}$$

The following functional is defined as

$$L_2\left(\Omega, \zeta_2\right) = -\frac{1}{m} \sum_{i=N_1}^{N_2} G_2(\boldsymbol{x}, \zeta_2), \tag{20}$$

where

$$G_2(\boldsymbol{x}, \zeta_2) = \boldsymbol{\alpha} \left\{ \omega \boldsymbol{I} \hat{\boldsymbol{u}} - \left(\int_{\Omega} \tilde{\boldsymbol{u}}^{\mathrm{T}} \tilde{\boldsymbol{u}} dx \right) \hat{\boldsymbol{u}} \right\}. \tag{21}$$

Actually, the index i is not needed in (21). After taking the Fréchet derivative for L_2, the index i is playing important role in (41).

3 Shape Optimization Problem

In this section, the new shape optimization problem using POD is constructed, where first the Lagrange function is defined to deduce the first variation. Next, based on the Kuhn–Tucker condition, the main and adjoint problems are solved to obtain the main and adjoint variables, which is substituted into the first variation to evaluate sensitivity for the shape optimization problem.

3.1 Lagrange Function and Its Material Derivative

We formulate the following a minimization problem of the cost function f as

$$f(\omega) = \frac{1}{m} \sum_{i=N_1}^{N_2} \omega_i. \tag{22}$$

Problem 3 (*Shape Optimization*) After letting $f(\omega)$ be defined as Eq. (22), find $\boldsymbol{\phi}(\Omega)$ such that

$$\min_{\boldsymbol{\phi}} \left\{ f(\omega) \; ; \; \left\{ (\boldsymbol{u}^n, p^n) \right\}_{n=N_1}^{N_2}, \; (\omega, \hat{\boldsymbol{u}}) \right\}. \tag{23}$$

See the description of (23) in [16].

By application of the Lagrange multiplier method, Lagrange function L for the shape optimization problem in this study is written as

$$L(\Omega, \zeta_1, \zeta_2) = f(\omega) + L_1(\Omega, \zeta_1) + L_2(\Omega, \zeta_2). \tag{24}$$

The first variation $\dot{L}$ of Lagrange function L with respect to arbitrary domain variation of $\boldsymbol{\varphi}$ is presented

$$\dot{L}(\Omega \cup \partial\Omega_{\mathrm{m}}, \zeta_1, \zeta_2, \boldsymbol{\phi}) = f'(\omega)$$
$$-\frac{1}{m}\sum_{n=N_1}^{N_2}\left\{\int_{\Omega} G_1'(\boldsymbol{x}, \zeta_1)\,dx + \int_{\partial\Omega_{\mathrm{m}}} G_1(\boldsymbol{x}, \zeta_1)\,\boldsymbol{v}\cdot\boldsymbol{\varphi}\,d\gamma\right\}$$
$$-\frac{1}{m}\sum_{i=N_1}^{N_2} G_2'(\boldsymbol{x}, \zeta_2), \tag{25}$$

where $f'(\omega)$ is the variation with respect to ω for the cost function f.

For simplicity in the next section, we defined the following functions as

$$\dot{L}(\Omega \cup \partial\Omega_{\mathrm{m}}, \zeta_1, \zeta_2, \boldsymbol{\phi})$$
$$= L_{\boldsymbol{w},q}(\Omega, \zeta_1) + L_{\boldsymbol{\alpha}}(\Omega, \zeta_2) + L_{\omega}(\Omega, \zeta_2) + L_{\hat{\boldsymbol{u}}}(\Omega, \zeta_2) + L_{\boldsymbol{u},p}(\Omega, \zeta_1)$$
$$-\int_{\partial\Omega_{\mathrm{m}}} G_1(\boldsymbol{x}, \zeta_1(\Omega))\,\boldsymbol{v}\cdot\boldsymbol{\varphi}\mathrm{d}\gamma, \tag{26}$$

where $L_{\boldsymbol{w},q}(\Omega, \zeta_1)$ and $L_{\boldsymbol{\alpha}}(\Omega, \zeta_2)$, $L_{\omega}(\Omega, \zeta_2)$, $L_{\hat{\boldsymbol{u}}}(\Omega, \zeta_2)$, $L_{\boldsymbol{u},p}(\Omega, \zeta_1)$ represent the Fréchet derivative with respect to $(\boldsymbol{w}, q)$ and $\boldsymbol{\alpha}$, ω, $\hat{\boldsymbol{u}}$, $(\boldsymbol{u}, p)$ as defined in Appendix B, and

$$-\frac{1}{m}\sum_{n=N_1}^{N_2}\left\{\int_{\Omega} G_1'(\boldsymbol{x}, \zeta_1)\,\mathrm{d}x\right\} = L_{\boldsymbol{u},p}(\Omega, \zeta_1) + L_{\boldsymbol{w},q}(\Omega, \zeta_1), \tag{27}$$

$$-\frac{1}{m}\sum_{i=N_1}^{N_2} G_2'(\boldsymbol{x}, \zeta_2) = L_{\omega}(\Omega, \zeta_2) + L_{\hat{\boldsymbol{u}}}(\Omega, \zeta_2) + L_{\boldsymbol{\alpha}}(\Omega, \zeta_2). \tag{28}$$

3.2 *Main and Adjoint Problems*

Based on adjoint variable method, stationary conditions known as the Kuhn–Tucker conditions in optimization theory are given as

$$L_{\boldsymbol{w},q}(\Omega, \zeta_1) = 0, \tag{29}$$
$$L_{\boldsymbol{\alpha}}(\Omega, \zeta_2 = 0, \tag{30}$$
$$L_{\omega}(\Omega, \zeta_2) = 0, \tag{31}$$
$$L_{\hat{\boldsymbol{u}}}(\Omega, \zeta_2) = 0, \tag{32}$$
$$L_{\boldsymbol{u},p}(\Omega, \zeta_1) = 0. \tag{33}$$

Equations (29) and (30), respectively, are the main problem and agree with Problem 1 and Problem 2. Equations (31)–(33) are the adjoint problems for ω, $\hat{\boldsymbol{u}}$, $(\boldsymbol{u}, p)$ in the shape optimization problem.

Problem 4 (*Adjoint Problem for* ω) After letting eigenfunction $\hat{\boldsymbol{u}}$ of Problem 2 be given, then find $\boldsymbol{\alpha} \in \mathbb{R}^{m\times m}$ such that

$$\hat{\boldsymbol{u}}\boldsymbol{\alpha} = \boldsymbol{I}, \tag{34}$$

and $\hat{\boldsymbol{u}}$, $\boldsymbol{\alpha}$ are the unitary matrix from Problem 4. Therefore $\boldsymbol{\alpha}$ is obtained as the inverse matrix or the transposed matrix of $\hat{\boldsymbol{u}}$,

$$\boldsymbol{\alpha} = \hat{\boldsymbol{u}}^{-1} = \hat{\boldsymbol{u}}^{\mathrm{T}}. \tag{35}$$

Problem 5 (*Adjoint Problem for* $\hat{\boldsymbol{u}}$) Let the solution $\boldsymbol{u}$ of Problem 1 be given with identify matrix $\boldsymbol{I}$. Find $\boldsymbol{\alpha}^{\mathrm{T}} \in \mathbb{R}^{m\times m}$ such that

$$\omega\boldsymbol{I}\boldsymbol{\alpha}^{\mathrm{T}} = \mathrm{R}(N_1, N_2, \tilde{\boldsymbol{u}}, \tilde{\boldsymbol{u}})\boldsymbol{\alpha}^{\mathrm{T}}. \tag{36}$$

In fact, it is not necessary to solve Problem 5 because $\boldsymbol{\alpha}$ has already been obtained in Problem 4.

From Appendix A and the time averaging of the non-stationary Navier–Stokes problem, Eq. (33), is rewritten as

$$\begin{aligned} L_{\boldsymbol{u},p}(\Omega, \zeta_1) = \\ &-\int_\Omega \{((\bar{\boldsymbol{u}}'\cdot\nabla)\,\bar{\boldsymbol{u}}) + ((\bar{\boldsymbol{u}}\cdot\nabla)\,\bar{\boldsymbol{u}}')\cdot\bar{\boldsymbol{w}} \\ &-\bar{p}'\nabla\cdot\bar{\boldsymbol{w}} - \bar{q}\nabla\cdot\bar{\boldsymbol{u}}' + \frac{1}{\mathrm{Re}}(\nabla\bar{\boldsymbol{u}}'^{\mathrm{T}}) : (\nabla\bar{\boldsymbol{w}}^{\mathrm{T}})\Big\}\,\mathrm{d}x, \\ &-\frac{1}{m}\sum_{i=N_1}^{N_2}\left\{-2\boldsymbol{\alpha}\left(\int_\Omega \tilde{\boldsymbol{u}}'^{\mathrm{T}}\tilde{\boldsymbol{u}}dx\right)\hat{\boldsymbol{u}}\right\}. \end{aligned} \tag{37}$$

All functions and function spaces are defined in Appendix E. The strong form of $L_{\boldsymbol{u},p}(\Omega, \zeta_1) = 0$ is derived using the following expressions as

$$\int_\Omega \{(\bar{\boldsymbol{u}}'\cdot\nabla)\bar{\boldsymbol{u}} + (\bar{\boldsymbol{u}}\cdot\nabla)\bar{\boldsymbol{u}}'\}\cdot\bar{\boldsymbol{w}}\mathrm{d}x = \int_\Omega \{(\nabla\bar{\boldsymbol{u}}^{\mathrm{T}})\bar{\boldsymbol{w}} - (\bar{\boldsymbol{u}}\cdot\nabla)\bar{\boldsymbol{w}}\}\cdot\boldsymbol{u}'\mathrm{d}x, \tag{38}$$

where

$$\int_\Omega \{\bar{\boldsymbol{u}}'\cdot\nabla)\bar{\boldsymbol{u}}\}\cdot\bar{\boldsymbol{w}} = \int_\Omega \{(\nabla\bar{\boldsymbol{u}}^{\mathrm{T}})\bar{\boldsymbol{w}}\}\cdot\bar{\boldsymbol{u}}'\mathrm{d}x, \tag{39}$$

and, based on $\nabla \cdot \boldsymbol{u} = 0$ in Ω,

$$\begin{aligned}\int_{\Omega} \left\{(\bar{\boldsymbol{u}} \cdot \nabla)\bar{\boldsymbol{u}}'\right\} \cdot \bar{\boldsymbol{w}} \mathrm{d}x &= \int_{\Omega} \nabla \cdot \left\{\bar{\boldsymbol{u}}(\bar{\boldsymbol{w}} \cdot \bar{\boldsymbol{u}}')\right\} \mathrm{d}x - \int_{\Omega} \{\nabla(\bar{\boldsymbol{u}}\bar{\boldsymbol{w}})\} \cdot \bar{\boldsymbol{u}}' \mathrm{d}x \\ &= -\int_{\Omega} \{\nabla(\bar{\boldsymbol{u}}\bar{\boldsymbol{w}})\} \cdot \bar{\boldsymbol{u}}' \mathrm{d}x \\ &= -\int_{\Omega} \{(\bar{\boldsymbol{u}} \cdot \nabla)\bar{\boldsymbol{w}} + \bar{\boldsymbol{w}}(\nabla \cdot \bar{\boldsymbol{u}})\} \cdot \bar{\boldsymbol{u}}' \mathrm{d}x \\ &= -\int_{\Omega} \{(\bar{\boldsymbol{u}} \cdot \nabla)\bar{\boldsymbol{w}}\} \cdot \bar{\boldsymbol{u}}' \mathrm{d}x. \end{aligned} \tag{40}$$

Furthermore, setting $\boldsymbol{\Phi}_{\omega} = \boldsymbol{\Phi}\omega^{-\frac{1}{2}}$,

$$\begin{aligned} -\frac{2}{m}\sum_{i=N_1}^{N_2}\left\{\boldsymbol{\alpha}\left(\int_{\Omega} \tilde{\boldsymbol{u}}'^{\mathrm{T}}\tilde{\boldsymbol{u}} \mathrm{d}x\right)\hat{\boldsymbol{u}}\right\} &= -\frac{2}{m}\sum_{i=N_1}^{N_2}\left(\int_{\Omega} \boldsymbol{\alpha}\tilde{\boldsymbol{u}}'^{\mathrm{T}}\tilde{\boldsymbol{u}}\hat{\boldsymbol{u}} \mathrm{d}x\right) \\ &= -\frac{2}{m}\sum_{i=N_1}^{N_2}\left(\int_{\Omega} \boldsymbol{\alpha}\tilde{\boldsymbol{u}}'^{\mathrm{T}}\boldsymbol{\Phi}\omega^{\frac{1}{2}} \mathrm{d}x\right) \\ &= -\frac{2}{m}\sum_{i=N_1}^{N_2}\left(\int_{\Omega} \boldsymbol{\alpha}\tilde{\boldsymbol{u}}'^{\mathrm{T}}\boldsymbol{\Phi}_{\omega} \mathrm{d}x\right) \\ &= -\frac{2}{m}\sum_{i=N_1}^{N_2}\left(\int_{\Omega} \boldsymbol{\alpha}\boldsymbol{\Phi}_{\omega}^{\mathrm{T}}\tilde{\boldsymbol{u}}' \mathrm{d}x\right) \\ &= -\int_{\Omega} \bar{\boldsymbol{A}} \cdot \bar{\boldsymbol{u}}' \mathrm{d}x, \end{aligned} \tag{41}$$

where $\bar{\boldsymbol{A}}$ and $\bar{\boldsymbol{u}}$ are calculated using

$$\bar{\boldsymbol{A}} = \frac{2}{m}\sum_{i=N_1}^{N_2}\boldsymbol{\alpha}\boldsymbol{\Phi}_{\omega}^{\mathrm{T}}, \ \bar{\boldsymbol{u}}' = \frac{1}{m}\sum_{n=N_1}^{N_2}\boldsymbol{u}'^{n}. \tag{42}$$

As a result, the adjoint problem for the time-averaged velocity and pressure $(\bar{\boldsymbol{u}}, \bar{p})$ is defined as shown below.

Problem 6 (*Adjoint Problem for* $(\bar{\boldsymbol{u}}, \bar{p})$) Let $\boldsymbol{\phi}$ and the time-averaged solution $(\bar{\boldsymbol{u}}, \bar{p})$ of Problem 1 and the eigenvalue and the eigenfunction $(\boldsymbol{\omega}, \hat{\boldsymbol{u}})$ of Problem 2 with $\hat{\boldsymbol{u}} = \boldsymbol{\alpha}^{\mathrm{T}}$ be given. Find $(\bar{\boldsymbol{w}}, \bar{q}) : \Omega \to \mathbb{R}^d \times \mathbb{R}$ such that

$$(\nabla \bar{\boldsymbol{u}}^{\mathrm{T}})\bar{\boldsymbol{w}} - (\bar{\boldsymbol{u}} \cdot \nabla)\bar{\boldsymbol{w}} + \nabla \bar{q} - \frac{1}{\mathrm{Re}}\Delta \bar{\boldsymbol{w}} + \bar{\boldsymbol{A}} = \boldsymbol{0} \text{ in } \Omega, \tag{43}$$

$$\nabla \cdot \bar{\boldsymbol{w}} = \boldsymbol{0} \text{ in } \Omega, \tag{44}$$

$$\bar{\boldsymbol{w}} = \boldsymbol{0} \text{ on } \partial\Omega. \tag{45}$$

3.3 Sensitivity of the Shape Optimization Problem

Here, substituting the solutions $\zeta_1 = \{\boldsymbol{u}, p, \boldsymbol{w}, q\}$ and $\zeta_2 = \left\{\omega, \hat{\boldsymbol{u}}, \boldsymbol{\alpha}, \boldsymbol{u}\right\}$ into $\dot{L}(\Omega \cup \partial\Omega_{\mathrm{m}}, \zeta_1, \zeta_2, \boldsymbol{\phi})$, we evaluate

$$\begin{aligned}
\dot{L}(\Omega \cup \partial\Omega_{\mathrm{m}}, \zeta_1, \zeta_2, \boldsymbol{\phi}) &= -\frac{1}{m}\sum_{n=N_1}^{N_2}\left(\int_{\partial\Omega_{\mathrm{m}}} G_1\left(\boldsymbol{x}, \zeta_1(\Omega)\right)\boldsymbol{v} \cdot \boldsymbol{\varphi}\mathrm{d}\gamma\right) \\
&= \int_{\partial\Omega_{\mathrm{m}}}\left\{-\frac{1}{m}\sum_{n=N_1}^{N_2} G_1\left(\boldsymbol{x}, \zeta_1(\Omega)\right)\right\}\boldsymbol{v} \cdot \boldsymbol{\varphi}\mathrm{d}\gamma \\
&= \int_{\partial\Omega_{\mathrm{m}}} \bar{G}_1\left(\boldsymbol{x}, \zeta_1(\Omega)\right)\boldsymbol{v} \cdot \boldsymbol{\varphi}\mathrm{d}\gamma
\end{aligned},$$

where $\bar{G}_1\left(\boldsymbol{x}, \zeta_1(\Omega)\right)$ is the sensitivity of the shape optimization problem,

$$\begin{aligned}
\bar{G}_1\left(\boldsymbol{x}, \zeta_1(\Omega)\right) &= -\frac{1}{m}\sum_{n=N_1}^{N_2}\frac{1}{\mathrm{Re}}\left\{\nabla \boldsymbol{u}^n\right\}^{\mathrm{T}} : \left\{\nabla \boldsymbol{w}^n\right\}^{\mathrm{T}} \\
&= -\frac{1}{\mathrm{Re}}\nabla \bar{\boldsymbol{u}}^{\mathrm{T}} : \nabla \bar{\boldsymbol{w}}^{\mathrm{T}}.
\end{aligned}$$

4 Numerical Schemes

The Taylor–Hood (P2-P1) element pair for the velocity and pressure is used to discretize all the equations spatially. FreeFEM++ [17] is used for all numerical calculations.

The stationary solution $(\boldsymbol{u}^0, p^0)$ is obtained to solve the stationary Navier–Stokes problem using the Newton–Raphson method. The non-stationary solution $\{(\boldsymbol{u}^n, p^n)\}_{n=1}^{N}$ is obtained to solve Problem 1 with UMFPACK [18] every time steps from $n = 1$ to N and the characteristic curve method [15].

After obtaining the non-stationary solution $\{(\boldsymbol{u}^n, p^n)\}_{n=1}^{N}$, the correlation coefficient matrix R is formed for snapshot POD. The eigenvalue problem for the matrix R is solved in Problem 2 using lapack solver.

Based on the theory of the optimization problem considered herein, the adjoint problem of Problem 6 is solved to obtain $(\bar{w}, \bar{q})$ with UMFPACK [18]. After evaluating the sensitivity, for domain deformation, the H^1 gradient method is used with UMFPACK [18].

5 Numerical Calculations and Discussion

5.1 Spatial and Temporal Discretization of the Optimization Problem

Velocity and pressure are discretized in the spatial direction by finite element method, where nodes and elements are $(N_{\text{nodes}}, N_{\text{elements}}) = (21945, 43290)$. For discretization in time, the time step size $\Delta t = 0.001$ is used to take time integrations of Problem 1. Velocity vectors are sampled from $T_1 = 3$ to $T_2 = 6$ for snapshot POD. Appendices C and D present more details of numerical accuracy and validations of sampled velocity fields for snapshot POD.

5.2 Numerical Results

In the initial domain, Fig. 1 shows stream functions of POD basis $\boldsymbol{\Phi}^i$ at the i primary components from $i = 1$ to 4. Figure 2 presents normalized cost function f^k/f^0 for reshaping step $k = 45$ at Re $= 100$, where the terminal condition $\beta = 10^{-6}$ and the small parameter $\epsilon \in [10^{-5}, 0.5]$ for the iteration scheme are shown in Sect. 2.3. In the optimal domain (Fig. 3), Fig. 4 shows the stream functions of the POD basis at the i primary components from $i = 1$ to 4.

Figure 2 shows the cost function with shape reshaping steps, and the values seems to converge sufficiently. Table 1 depicts eigenvalues and the power spectral density obtained from snapshot POD in the initial and the optimal domains. From Table 1, the power spectrum density is over 95% up to the second primary component in both domains.

The result reveals that the cost function (sum of the eigenvalues) is decreasing. Especially the eigenvalues except the first primary component are decreasing. Therefore, the amplitude of the time periodic flow in the optimal domain is suppressed more than them in the initial domain. However, the eigenvalue for the first primary component is increasing. The time-averaged flow is developing.

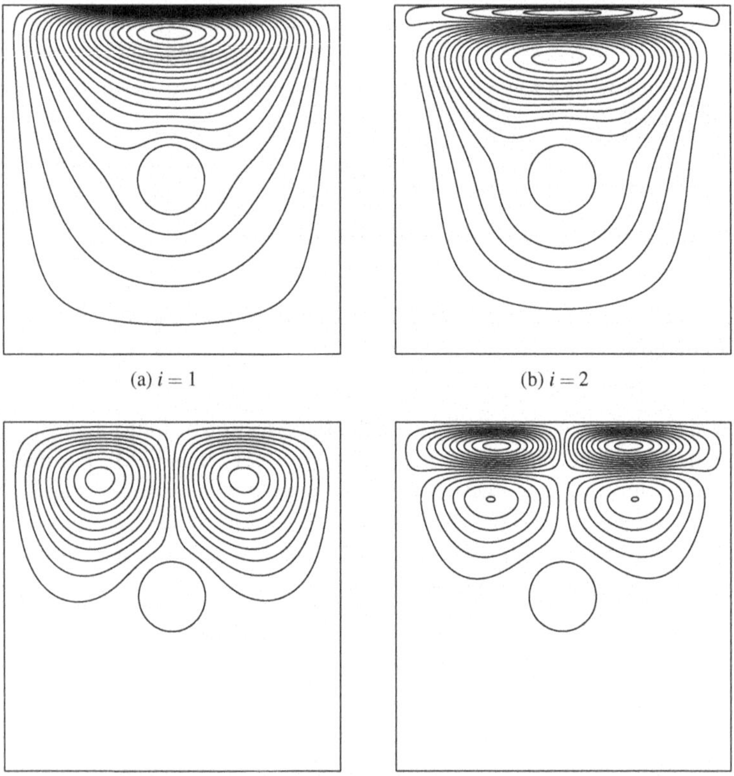

(a) $i = 1$ (b) $i = 2$

(a) $i = 3$ (b) $i = 4$

Fig. 1 Stream functions of POD basis at ith primary components from $i = 1$ to $i = 4$ at Re $= 100$ in the initial domain Ω

Fig. 2 Normalized cost function f^k/f^0 for reshaping step $k = 45$ at Re $= 100$

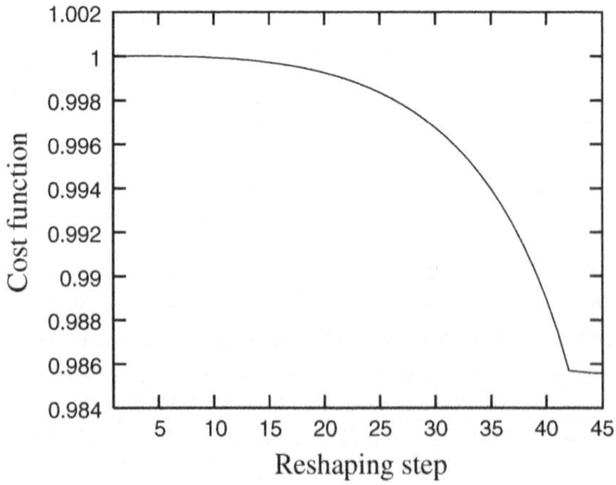

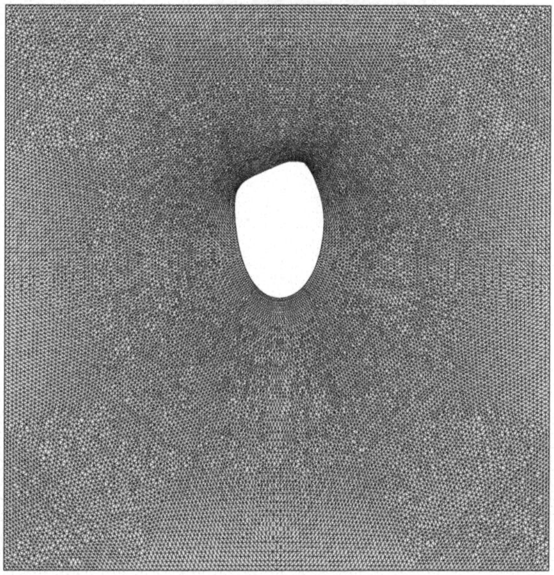

Fig. 3 The optimal domain

Table 1 At ith primary components, eigenvalues ω_i and the power spectral density, up to $i = 4$ in the initial and the optimal domains

i	Initial domain (Power spectral density)	Optimal domain (Power spectral density)
1	0.151843(83.69%)	0.152228(85.13%)
2	0.022432(96.05%)	0.021392(97.09%)
3	0.006833(99.82%)	0.004869(99.82%)
4	0.000233(99.95%)	0.000059(99.95%)

6 Conclusions

As described in this paper, the author formulates a new shape optimization method for suppressing time periodic flow fields driven only by the non-stationary boundary condition at a sufficiently low Reynolds number considering snapshot Proper Orthogonal Decomposition (POD). Particularly the sum of eigenvalues in POD is defined as the cost function. The non-stationary Navier–Stokes problem and the eigenvalue problem in POD are used as main problems. The main problems are transformed from strong forms to weak forms with trial functions based on a standard framework of the Finite Element Method (FEM). The functional is described using the Lagrange multiplier method with FEM. Next, its material derivative is derived to evaluate the sensitivity using adjoint variable method. The initial domain is reshaped iteratively,

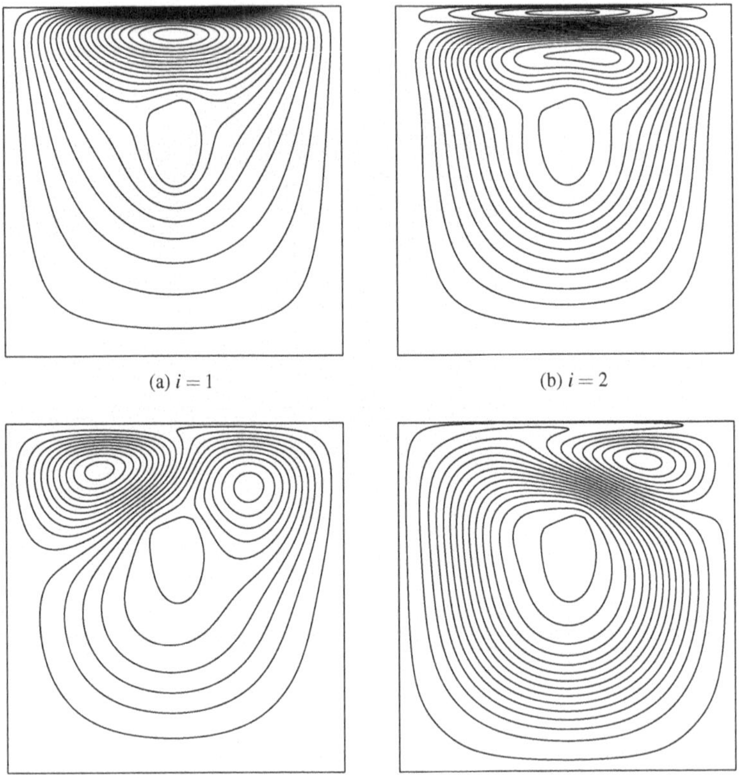

(a) $i = 1$ (b) $i = 2$

(a) $i = 3$ (b) $i = 4$

Fig. 4 Stream functions of POD basis at ith primary components from $i = 1$ to $i = 4$ at Re = 100 in the optimal domain $\boldsymbol{\phi}(\Omega)$

where the H^1 gradient method is used for stable domain deformation. For a numerical demonstration, a two-dimensional cavity flow with a disk-shaped isolated body is used.

Appendix A: Variational Form of the Non-stationary Navier–Stokes Problem

To discretize in time, we use the characteristic finite element scheme, where Δt is the time step satisfying the total number of time steps $N = T/\Delta t$, and the time at $n \in \mathbb{N}$ step is denoted by $t^n = n\Delta t$ for $n \in \{0 \ldots, N\}$. $\boldsymbol{x}$ and $\boldsymbol{X}^n$, respectively, denote the positions at time t^{n+1} and t^n of the particle. More details of the finite element method are explained in Sect. 2 of Notsu [15].

Finally, we deduce the following weak formulations using the characteristic finite element scheme to find $\{(\boldsymbol{u}^n, p^n)\}_{n=1}^N$ such that for $n \in \{0 \ldots, N\}$

$$-\int_{\Omega} \left\{ \frac{\boldsymbol{u}^{n+1}(\boldsymbol{x}) - \boldsymbol{u}^n(\boldsymbol{X}^n)}{\Delta t} \cdot \boldsymbol{w}^{n+1} - p^{n+1}\nabla \cdot \boldsymbol{w}^{n+1} - q^{n+1}\nabla \cdot \boldsymbol{u}^{n+1} \right.$$
$$\left. + \frac{1}{\mathrm{Re}}(\nabla \boldsymbol{u}^{n+1})^{\mathrm{T}} : (\nabla \boldsymbol{w}^{n+1})^{\mathrm{T}} \right\} \mathrm{d}x = 0, \tag{46}$$

for all $\{(\boldsymbol{w}^n, q^n)\}_{n=1}^N$ with the initial value $\boldsymbol{u}^0$, where all functions and function spaces are defined in Appendix E. Hereinafter, for convenience, $\boldsymbol{u} = \{\boldsymbol{u}^n\}_{n=0}^N$, and $p = \{p^n\}_{n=0}^N$, $\boldsymbol{w} = \{\boldsymbol{w}^n\}_{n=0}^N$, $q = \{q^n\}_{n=0}^N$.

After solving Problem 1, we obtain the time-averaged velocity and pressure $(\bar{\boldsymbol{u}}, \bar{p})$ by taking the time average for $\{(\boldsymbol{u}^n, p^n)\}_{n=N_1}^{N_2}$

$$\bar{\boldsymbol{u}} = \frac{1}{m}\sum_{n=N_1}^{N_2} \boldsymbol{u}^n, \; \bar{p} = \frac{1}{m}\sum_{n=N_1}^{N_2} p^n,$$

for $m = N_2 - N_1 + 1$. Time-averaged velocity and pressure are governed by the time-averaged Navier–Stokes problem.

The weak forms of the time-averaged Navier–Stokes problem are written as

$$-\int_{\Omega} \left\{ ((\bar{\boldsymbol{u}} \cdot \nabla)\bar{\boldsymbol{u}}) \cdot \bar{\boldsymbol{w}} - \bar{p}\nabla \cdot \bar{\boldsymbol{w}} - \bar{q}\nabla \cdot \bar{\boldsymbol{u}} + \frac{1}{\mathrm{Re}}(\nabla \bar{\boldsymbol{u}}^{\mathrm{T}}) : (\nabla \bar{\boldsymbol{w}}^{\mathrm{T}}) \right\} \mathrm{d}x = 0, \tag{47}$$

for all $(\bar{\boldsymbol{w}}, \bar{q})$.

For the optimization problem, let $\zeta_1 = \{\boldsymbol{u}, p, \boldsymbol{w}, q\}$ represent the set of necessary variables used in Problem 1. For $\Delta t \to 0$ we assume the following functional as

$$L_1(\Omega, \zeta_1) = -\frac{1}{m}\sum_{n=N_1}^{N_2} \left\{ \int_{\Omega} G_1(\boldsymbol{x}, \zeta_1)\,\mathrm{d}x \right\}$$
$$= -\int_{\Omega} \left\{ ((\bar{\boldsymbol{u}} \cdot \nabla)\bar{\boldsymbol{u}}) \cdot \bar{\boldsymbol{w}} - \bar{p}\nabla \cdot \bar{\boldsymbol{w}} - \bar{q}\nabla \cdot \bar{\boldsymbol{u}} + \frac{1}{\mathrm{Re}}(\nabla \bar{\boldsymbol{u}}^{\mathrm{T}}) : (\nabla \bar{\boldsymbol{w}}^{\mathrm{T}}) \right\} \mathrm{d}x.$$

where the density function for L_1 is

$$G_1(\boldsymbol{x}, \zeta_1) = \frac{\boldsymbol{u}^{n+1}(\boldsymbol{x}) - \boldsymbol{u}^n(\boldsymbol{X}^n)}{\Delta t} \cdot \boldsymbol{w}^{n+1}$$
$$-p^{n+1}\nabla \cdot \boldsymbol{w}^{n+1} - q^{n+1}\nabla \cdot \boldsymbol{u}^{n+1} + \frac{1}{\mathrm{Re}}(\nabla \boldsymbol{u}^{n+1})^{\mathrm{T}} : (\nabla \boldsymbol{w}^{n+1})^{\mathrm{T}}.$$

Appendix B: Variational Forms for the Main and the Adjoint Problems

Particularly based on the chain rule, we have

$$L_{w,q}(\Omega, \zeta_2) = -\frac{1}{m}\sum_{n=N_1}^{N_2}\left[\int_{\Omega}\left\{\frac{u^{n+1}(x) - u^{n}(X^{n})}{\Delta t}\cdot w'^{n+1}\right.\right.$$

$$\left.\left. -p^{n+1}\nabla\cdot w'^{n+1} - q'^{n+1}\nabla\cdot u^{n+1} + \frac{1}{\mathrm{Re}}(\nabla u^{n+1})^{\mathrm{T}} : (\nabla w'^{n+1})^{\mathrm{T}}\right\}\mathrm{d}x\right],$$

$$L_{\alpha}(\Omega, \zeta_2) = -\frac{1}{m}\sum_{i=N_1}^{N_2}\alpha'\left\{\omega I\hat{u} - \left(\int_{\Omega}\tilde{u}^{\mathrm{T}}\tilde{u}\mathrm{d}x\right)\hat{u}\right\},$$

$$L_{\omega}(\Omega, \zeta_1) = \frac{1}{m}\sum_{i=N_1}^{N_2}\left\{\omega'_i\left(I - \alpha\hat{u}\right)\right\},$$

$$L_{\hat{u}}(\Omega, \zeta_2) = -\frac{1}{m}\sum_{i=N_1}^{N_2}\left\{\alpha\left\{\omega I\hat{u}' - \left(\int_{\Omega}\tilde{u}^{\mathrm{T}}\tilde{u}dx\right)\hat{u}'\right\}\right\},$$

$$L_{u,p}(\Omega, \zeta_1) = -\frac{1}{m}\sum_{n=N_1}^{N_2}\left[\int_{\Omega}\left\{\frac{u'^{n+1}(x) - u'^{n}(X^{n})}{\Delta t}\cdot w^{n+1}\right.\right.$$

$$\left.\left. -p'^{n+1}\nabla\cdot w^{n+1} - q^{n+1}\nabla\cdot u'^{n+1} + \frac{1}{\mathrm{Re}}(\nabla u'^{n+1})^{\mathrm{T}} : (\nabla w^{n+1})^{\mathrm{T}}\right\}\mathrm{d}x\right]$$

$$-\frac{1}{m}\sum_{i=N_1}^{N_2}\left\{-2\alpha\left(\int_{\Omega}\tilde{u}'^{\mathrm{T}}\tilde{u}\mathrm{d}x\right)\hat{u}\right\}.$$

Appendix C: Numerical Accuracy

For ensuring numerical accuracy, validation of discretization in the spatial and the time directions are operated by solving Problem 3 at $\mathrm{Re} = 100$, where velocity vectors are sampled from $T_1 = 3$ to $T_2 = 6$ for POD. First, $\Delta t = 0.001$ is fixed. The combination of nodes and elements $(N_{\text{nodes}}, N_{\text{elements}})$ represented in Table 2 is selected. The eigenvalue ω^2 for the second primary component in the optimal domain is chosen as the index to compare the numerical accuracy. As a result, ω^2 converges for the combination of nodes and elements (21945, 43290).

Second, $(N_{\text{nodes}}, N_{\text{elements}}) = (21945, 43290)$ is fixed. Δt is changed from 0.05 to 0.001. Again ω^2 in the optimal domain is chosen as the index for numerical accuracy. From Table 3, ω^2 converges for $\Delta t = 0.001$.

Table 2 Seven cases of $(N_{\text{nodes}}, N_{\text{elements}})$ and the eigenvalue ω^2 for the second primary component in the optimal domain under $\Delta t = 0.001$ and $T_2 = 6$

$(N_{\text{nodes}}, N_{\text{elements}})$, Δt, T_2	ω^2
(1399, 2648), 0.001, 6	0.02172
(5477, 10654), 0.001, 6	0.02161
(12129, 23808), 0.001, 6	0.02132
(21945, 43290), 0.001, 6	0.02120
(21945, 43290), 0.001, 6	0.02139
(25264, 49878), 0.001, 6	0.02138
(29433, 58166), 0.001, 6	0.02137

Table 3 Four cases of Δt and the eigenvalue ω^2 for the second primary component in the optimal domain, under the combination of nodes and elements (21945, 43290) and $T_2 = 6$

$(N_{\text{nodes}}, N_{\text{elements}})$, Δt, T_2	ω^2
(21945, 43290), 0.05, 6	0.01672
(21945, 43290), 0.01, 6	0.02045
(21945, 43290), 0.005, 6	0.02096
(21945, 43290), 0.001, 6	0.02139

Appendix D: Validations of Sampled Velocity Fields for Snapshot POD

Fixed at $(N_{\text{nodes}}, N_{\text{elements}}) = (21945, 43290)$, and $\Delta t = 0.001$, Problem 3 is solved at $\text{Re} = 100$, where velocity vectors are sampled from $T_1 = 3$ to T_2 represented in Table 4 for POD. From Table 4 shows that at ω^2 is converging at $T_3 = 6$ sufficiently.

Table 4 Eight cases of T_2 and $\frac{1}{(T_2-T_1)/\Delta t}\omega^2$ in the optimal domain, fixed at $(N_{\text{nodes}}, N_{\text{elements}}) =$ (21945, 43290) and $\Delta t = 0.001$, $T_1 = 3$

$(N_{\text{nodes}}, N_{\text{elements}})$,$\Delta t$,$T_2$	$\frac{1}{(T_2-T_1)/\Delta t}\omega^2$
(21945, 43290), 0.001, 3.25	0.00418
(21945, 43290), 0.001, 3.5	0.01079
(21945, 43290), 0.001, 3.75	0.00768
(21945, 43290), 0.001, 4.0	0.00768
(21945, 43290), 0.001, 4.5	0.00799
(21945, 43290), 0.001, 5.0	0.00745
(21945, 43290), 0.001, 5.5	0.00761
(21945, 43290), 0.001, 6.0	0.00713

However, ω^2 is converging over $T_2 = 4$ already and a period of the oscillating top boundary condition is the non-dimensional time 1. Therefore, Appendix D shows numerically that a secondary flow is not developed.

Appendix E: Functions and Function Spaces

A set of velocity vector $\boldsymbol{u}^n \in U$ and pressure $p^n \in P$ at every time steps n are introduced as

$$U = \left\{ \boldsymbol{u} \in H^1(\boldsymbol{\phi}(\Omega); \mathbb{R}^2) \mid \boldsymbol{u} = \boldsymbol{u}_{\mathrm{D}} \cos(2\pi n \Delta t) \text{ on } \boldsymbol{\phi}(\partial\Omega) \right\},$$

$$P = \left\{ p \in L^2(\boldsymbol{\phi}(\Omega); \mathbb{R}) \mid \int_{\boldsymbol{\phi}(\Omega)} p dx = 0 \right\}.$$

A set of Lagrange multiplier $(\boldsymbol{w}^n, q^n) \in W \times Q$ for $(\boldsymbol{u}^n, p^n) \in U \times P$ are introduced as

$$W = \left\{ \boldsymbol{w} \in H^1(\boldsymbol{\phi}(\Omega); \mathbb{R}^2) \mid \boldsymbol{w} = \boldsymbol{0} \text{ on } \boldsymbol{\phi}(\partial\Omega) \right\},$$
$$Q = P.$$

A set of time-average velocity and pressure, its Lagrange multiplier are $(\bar{\boldsymbol{u}}, \bar{p}) \in W \times P$ and $(\bar{\boldsymbol{w}}, \bar{q}) \in W \times Q$, and more the Fréchet derivative for time-average velocity and pressure, its Lagrange multiplier are $(\bar{\boldsymbol{u}}', \bar{p}') \in W \times P$ and $(\bar{\boldsymbol{w}}', \bar{q}') \in W \times Q$.

Funding Information This study was funded by JSPS KAKENHI (16K20906).

Compliance with Ethical Standards.

Conflict of Interest The authors declare that they have no conflict of interest.

References

1. Nakazawa, T., Azegami, H.: Shape Optimization Method improving Hydrodynamic Stability. Jpn. J. Ind. Appl. Math. **33**, 167–181 (2015)
2. Haslinger, J., MakinenR, A.E.: Introduction to Shape Optimization: Theory, Approximation, and Computation. SIAM, Philadelphia (2003)
3. Mohammadi, B., Pironneau, O.: Applied Shape Optimization for Fluids. Oxford University Press, Oxford (2001)
4. Moubachir, M., Zolesio, J.P.: Moving Shape Analysis and Control: Applications to Fluid Structure Interactions. Chapman and Hall/CRC Pure and Applied Mathematics, Boca Raton (2006)
5. Shinohara, K., Okuda, H., Ito, S., Nakajima, N., Ida, M.: Shape optimization using adjoint variable method for reducing drag in Stokes flow. Int. J. Numer. Meth. Fluids **58**, 119–159 (2008)

6. Katamine, E., Nagatomo, Y., Azegami, H.: Shape optimization of 3D viscous flow fields. Inverse Prob. Sci. Eng. **17**, 105–114 (2009)
7. Ghosh, S., Pratihar, D.K., Maiti, B., Das, P.K.: An evolutionary optimization of diffuser shapes based on CFD simulations. Int. J. Numer. Meth. Fluids **63**, 1147–1166 (2010)
8. Iwata, Y., Azegami, H., Katamine, E.: Numerical Solution to shape optimization problem for non-stationary Navier-Stokes problems. JSIAM Lett. **2**, 37–40 (2010)
9. Li, D., Hartman, R.: Adjoint-based airfoil optimization with discretization error control. Int. J. Numer. Meth. Fluids. **77**, 1–17 (2015)
10. Schmidt, S., Ilic, C., Schulz, V., Gauger, N.: Three-dimensional large-scale aerodynamic shape optimization based on shape calculus. AIAA J. **51**, 2615–2627 (2013)
11. Nakazawa, T.: Increasing the Critical Reynolds number by maximizing dissipation energy problem. In: Proceedings of the Fifth International Conference on Jets, Wakes and Separated Flows (ICJWSF2015), Editor Antonio Segalini, Chap. 75 (2016)
12. Azegami, H., Wu, Z.: Domain optimization analysis in linear elastic problems: approach using traction method. JSME Int. J. Ser. A **39**, 272–278 (1996)
13. Imam, H.M.: Three-dimensional shape optimization. Int. J. Num. Methods Eng. **18**, 661–673 (1982)
14. Nakazawa, T.: The two step shape optimization algorithm improving hydrodynamics stability. Proc. ECCOMAS Congress **2016**, 4881 (2016)
15. Notsu, H., Tabata, M.: Error estimates of a pressure-stabilized characteristics finite element scheme for Oseen equations. J. Sci. Comput. **65**, 940–955 (2015)
16. Azegami, H., Solution of shape optimization problem and its application to product design. In: Van Meurs, P., Kimura, M., Notsu, H. (eds.) Mathematical Analysis of Continuum Mechanics and Industrial Applications II (Proc. CoMFoS16), pp. 83–98. Springer, Berlin (2017)
17. Hecht, F.: New development in FreeFem++. J. Numer Math. **20**, 251–265 (2012)
18. Davis, T.: Algorithm 832: UMFPACK, an unsymmetric-pattern multifrontal method. ACM Trans. Math. Softw. **30**, 196–199 (2004)
19. Kimura, K.: Lecture notes Volume IV. Top. Math. Model. **1–38** (2008)

Topology Optimization for Porous Cooling Systems

Kentaro Yaji

Abstract The aim of this paper is to investigate an applicability of topology optimization in the design of porous cooling systems. The basic concept of porous cooling systems is to utilize the combination of coolant flow channel and porous material for efficiently boosting heat transfer performances. For generating nonintuitive design candidates of porous cooling systems, we formulate a topology optimization problem, whose purpose is to determine optimum distributions with respect to fluid and permeable solid; in contrast, conventional formulations in topology optimization for the design of cooling systems focus on determining the distributions of fluid and impermeable solid. We provide a numerical example to demonstrate the efficacy of topology optimization in the design of porous cooling systems.

1 Introduction

Topology optimization [4, 5] is a mathematical optimization method, in which objective function is minimized/maximized via changing structural shape, and has been attracted attention as a powerful design tool for realizing innovative designs. Its attractive feature is that topology optimization can automatically generate topology-optimized configurations from a nonintuitive initial design domain, whereas sizing or shape optimizations—traditional structural optimizations—typically require a priori promising initial designs (see Fig. 1). Although topology optimization has been mainly applied to structural mechanics problems, various physical problems have been also treated in topology optimization problems, such as thermal, electromagnetic, and fluid problems. In addition, the applicability of topology optimization in practical engineering design problems has been actively investigated by academic and industrial researchers. The reader is referred to as pioneering monograph [5],

K. Yaji (✉)
Department of Mechanical Engineering, Graduate School of Engineering,
Osaka University, 2-1, Yamadaoka, Suita, Osaka 565-0871, Japan
e-mail: yaji@mech.eng.osaka-u.ac.jp

H. Itou et al. (eds.), *Mathematical Analysis of Continuum Mechanics and Industrial Applications III*, Mathematics for Industry 34,
https://doi.org/10.1007/978-981-15-6062-0_10

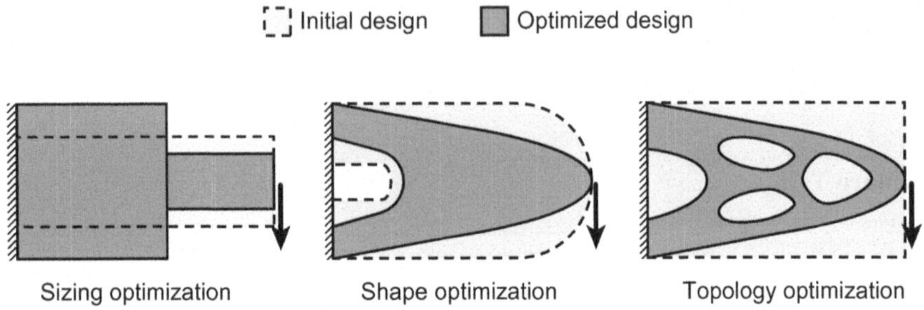

Fig. 1 Schematic illustration of structural optimization methods, in which the representative structural optimization problem—stiffness maximization problem—is shown

and recent developments in the research field of topology optimization can be seen in review articles [9, 22].

Topology optimization for fluid problems [6, 7, 11, 12, 15, 16, 18, 20] is relatively new compared with the cases dealing with other physics. This research topic was first discussed by Borravall and Petersson in 2003 [6], and many researchers have reported the applicability of the fluid topology optimization in the design of cooling systems such as heatsinks [1, 2, 8, 14, 19, 21, 27–29, 31]. Results in the previous works dealing with such cooling system design problems indicated that geometrically complex configurations are effective for boosting target performances. For instance, branched channel configurations, which resemble *a vein of a plant*, are often obtained as topology-optimized configurations for maximizing heat transfer efficiencies of micro-channel heatsinks. It can be safely said that such complex and interesting configurations cannot be generated by only designer's intuitions; furthermore, the optimized configurations generally achieve superior performances in comparison with existing simple configurations. Due to these attractive features, topology optimization holds great potential for realizing innovative designs of practical cooling systems.

In general, topology optimization for cooling system design problems is formulated as a material distribution problem with respect to fluid and solid in a design domain. For this, an artificial body force term based on the Brinkman approach is typically applied so that only the solid domain is expressed as zero-velocity field [6]. This idea is interpreted as considering that the solid domain is expressed as a very low permeability material. That is, the typical strategy for solving the cooling system design problems has focused on solving the material distribution problem with respect to fluid and "impermeable" solid.

In this paper, we discuss topology optimization formulated as a material distribution problem with respect to fluid and "permeable" solid. The permeable solid is interpreted as porous medium, which allows to pass through the fluid. Since porous medium has large surface area, which facilitates to boost the heat transfer efficiency, many researchers have proposed the so-called *porous cooling systems* such as porous heat sinks [10, 23, 26]. Thus, the aim of this paper is to investigate the applicability of topology optimization for the design of porous cooling systems. As far as we know, although there is no research focusing on such cooling system design problem in the research community of topology optimization, it should be noted that the material distribution problem over fluid and practical porous medium was already discussed in a previous work dealing with a flow battery design problem [30]. That is, our study is an application of this previous work for the design of porous cooling systems.

This paper is organized as follows. Section 2 provides governing equations for a thermal-fluid system. Section 3 describes the basic concept of topology optimization and formulates an optimization problem for the design of porous cooling systems. Section 4 presents our numerical result, and we conclude with Sect. 5.

2 Governing Equations

Consider a smooth bounded domain, $\Omega_\mathrm{f} \subset \mathbb{R}^d$ (d: spatial dimension), which is fulfilled with a steady-state incompressible viscous fluid. In the fluid domain Ω_f, velocity $\mathbf{u} : \Omega_\mathrm{f} \to \mathbb{R}^d$ and pressure $p : \Omega_\mathrm{f} \to \mathbb{R}$ in non-dimensional form are governed by the following equations:

$$\nabla \cdot \mathbf{u} = 0, \tag{1}$$

$$(\mathbf{u} \cdot \nabla)\mathbf{u} = -\nabla p + \frac{1}{Re}\nabla^2\mathbf{u} + \mathbf{F}. \tag{2}$$

Here, $\mathbf{F} : \Omega_\mathrm{f} \to \mathbb{R}^d$ is the body force, and $Re > 0$ is a non-dimensional parameter, the so-called Reynolds number, given by $\rho U L/\mu$, where ρ, U, L, and μ are the fluid density, a reference velocity speed, a reference length, and the fluid viscosity, respectively.

In addition, we consider forced convection heat transfer, which means temperature $T : \Omega_\mathrm{f} \to \mathbb{R}$ is dealt with passive scalar, expressed as follows:

$$Re\, Pr\, \mathbf{u} \cdot \nabla T = \nabla^2 T + Q, \tag{3}$$

where $Q : \Omega_\mathrm{f} \to \mathbb{R}$ is the volumetric heat source, and $Pr > 0$ is a non-dimensional parameter, called Prandtl number, given by $k/(c\rho)$, where k, and c are the thermal conductivity and the specific heat, respectively.

The above partial differential equations (1)–(3) can be solved under appropriate boundary conditions. Note that the governing equation (3) for T can be solved alone after solving the governing equations (1) and (2) for $(\mathbf{u}, p)$.

3 Topology Optimization for Thermal-Fluid Systems

3.1 Topology Optimization

The basic idea of topology optimization is to introduce a fixed design domain $D \subset \mathbb{R}^d$ that includes an original design domain, $\Omega \subseteq D$, and to express material distribution in D by using characteristic function $\chi \in L^\infty(D)$ defined as

$$\chi(\mathbf{x}) = \begin{cases} 1 & \text{if } \mathbf{x} \in \Omega \\ 0 & \text{if } \mathbf{x} \in D \setminus \Omega, \end{cases} \tag{4}$$

where $\forall \mathbf{x} \in D$ is the position.

According to the original concept of topology optimization, although χ should be directly controlled during the optimization process, the discrete nature of χ makes it difficult to solve. Hence, relaxation techniques have been proposed for efficiently solving topology optimization problems. Among them, one of the most popular approaches is the density approach [3], whose basic idea is to replace χ with a continuous one. The continuous design field γ belongs to the following set:

$$\mathscr{X}_\gamma = \left\{\gamma \in L^\infty(D) \,|\, 0 \leqslant \gamma(\mathbf{x}) \leqslant 1 \text{ a.e. } \mathbf{x} \text{ in } D\right\}. \tag{5}$$

The merit of using the continuous function is that gradient-based optimization algorithms can be exploited for efficiently searching optimized distribution of γ in D. On the other hand, it should be noted that intermediate region $0 < \gamma(\mathbf{x}) < 1$, called grayscale, is allowed during optimization process. Removing grayscale is essential from an engineering standpoint and has been widely studied in the research community of topology optimization [5, 22].

3.2 Expansion of Governing Equations for Thermal-Fluid Systems

We now discuss the design field γ in topology optimization problems dealing with thermal-fluid systems. The role of γ is to expand the governing equations (1)–(3), which are defined on Ω_{f}, for expressing both of fluid and solid domains $\Omega_{\mathrm{all}} = \Omega_{\mathrm{f}} \cup \Omega_{\mathrm{s}}$, where Ω_{s} is the solid domain. For this, we introduce an artificial body force in (2), as follows:

$$\mathbf{F} = -\alpha_\gamma \mathbf{u} \quad \text{with} \quad \alpha_\gamma = \frac{q(1-\gamma)}{\gamma + q}\alpha, \tag{6}$$

where $q > 0$ is a parameter for tuning the convexity of α_γ, and $\alpha \to \infty$ is the so-called inverse permeability [6] for expressing the solid domain Ω_s. Due to (6), the solid domain ($\gamma = 0$) can be expressed as $\mathbf{u} \to \mathbf{0}$, whereas the fluid domain ($\gamma = 1$) is not affected the artificial body force, namely, $\mathbf{F} = \mathbf{0}$. As a result, the expanded governing equations using (6) can express both of fluid and solid domains in the fixed design domain $D \subseteq \Omega_\mathrm{all}$. Note that α is typically set to a large value, e.g., 10^4, for its numerical treatment. However, it should be emphasized that practical permeability values of porous models can be selected as α. This suggests that topology optimization enables an innovative design of a cooling system composed of thermal-fluid and porous medium. We will discuss the details of porous model in Sect. 3.3.

Let us consider that heat source is applied to the solid domain Ω_s. We introduce an artificial heat source in (3), as follows:

$$Q = \beta_\gamma(1 - T) \quad \text{with} \quad \beta_\gamma = \beta(1 - \gamma), \tag{7}$$

where $\beta > 0$ denotes the volumetric heat transfer coefficient. As with the case of using (6), the expanded governing equation using (7) can express both of fluid and heated solid domains in the fixed design domain $D \subseteq \Omega_\mathrm{all}$.

3.3 Porous Model

As mentioned in Sect. 3.2, the inverse permeability can be selected as a practical one of the porous models. For instance, the Carman–Kozeny model [25] is given by

$$K = \frac{d_\mathrm{f}^2 \varepsilon^3}{16 k_\mathrm{CK}(1 - \varepsilon)^2}. \tag{8}$$

Here, K ($= \mu/\alpha$) is the dimensional permeability of porous medium, d_f is the fiber diameter, $\varepsilon \in [0, 1]$ is the porosity, and $k_\mathrm{CK} > 0$ is an empirical constant, called Carman–Kozeny constant, which depends on the type of porous medium.

By using such porous model, a cooling system composed of fluid and permeable solid can be designed on the basis of topology optimization for thermal-fluid systems, whereas the previous works [19, 28] focus on the cooling system composed of fluid and impermeable solid.

Note that the main research interest in this paper is to investigate the effect of α over the topology-optimized configuration in cooling system design problems. The concrete investigation is conducted through a numerical example in Sect. 4.

3.4 Formulation

According to the previous works [19, 28], a cooling system design problem can be formulated, on the basis of the density approach, as follows:

$$\begin{aligned} &\underset{\gamma(\mathbf{x})}{\text{maximize}} \;\; J = \int_D Q \, \mathrm{d}\Omega, \\ &\text{subject to} \;\; 0 \leqslant \gamma(\mathbf{x}) \leqslant 1 \quad \text{for } \forall \mathbf{x} \in D, \end{aligned} \tag{9}$$

where J is the objective functional, Q is the volumetric heat transfer defined as (7). In the optimization problem (9), the thermal-fluid system is governed by (1)–(3) with the artificial terms defined as (6) and (7).

The optimization problem (9) can be solved using a gradient-based optimization algorithm. For the implementation based on mathematical programming, the analysis domain Ω_{all} is discretized using finite number of elements, and the discretized design field—the design variables in mathematical programming—is allocated on each element, i.e., γ_e with the element number $e \in \mathbb{N}^+$. The gradient information, $\mathrm{d}J/\mathrm{d}\gamma_e$, is derived via adjoint method, which enables efficient gradient-based optimization without depending on the number of design variables. The reader is referred to the monograph [5] for the details of adjoint method.

4 Numerical Example

4.1 Details of Numerical Settings

In this paper, we used sequential linear programming (SLP) as a gradient-based optimization algorithm on the basis of the previous work [30]. The analysis domain shown in Fig. 2 was discretized using $N = 80,800$ square elements based on the finite element method (FEM), and the velocity $\mathbf{u}$, the pressure p, and the temperature T were discretized using bilinear finite elements. To satisfy the inf-sup conditions and guarantee stability, the streamline-upwind/Petrov–Galerkin (SUPG) and pressure-stabilized/Petrov–Galerkin (PSPG) stabilization techniques were employed [24]. The non-dimensional parameters for solving the thermal-fluid system were set so that $Re = 1.2 \times 10^3$, $Pr = 7.0$, and $\beta = 1.0 \times 10^2$.

The initial design variables were set to $\gamma_e = 0.99$ ($e = 1, \ldots, N$), which means the fixed design domain D is almost fulfilled by fluid. On the basis of previous work [6], the initial tuning parameter for α_γ was set to $q = 1.0 \times 10^{-2}$, which was gradually increased until 1.0 during optimization process, for removing grayscale. In addition, a partial differential equation (PDE)-based filter [13, 17] was used for ensuring smoothness of the design variables in D during the optimization process.

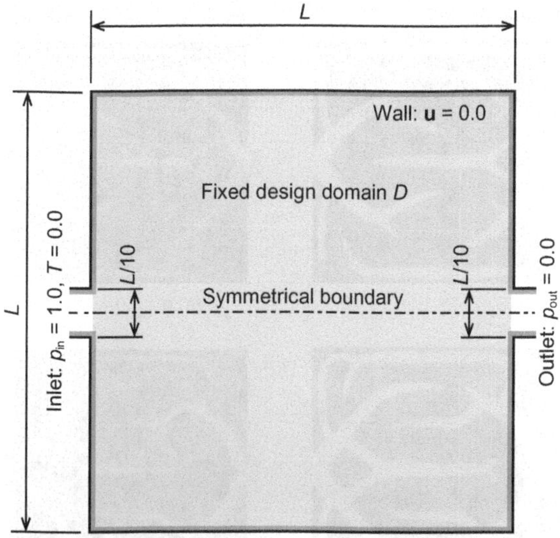

Fig. 2 Analysis domain for the cooling system design problem

4.2 Effect of α over Topology-Optimized Configuration

We investigated the effect of α over the topology-optimized configuration. Figure 3 shows the obtained results, where $\alpha = 2.0 \times 10^2, 5.0 \times 10^2, 1.0 \times 10^3, 1.0 \times 10^4$ were, respectively, set so that various patterns of topology-optimized configurations can be derived.

As shown in Fig. 3, the disconnected flow channels are observed when setting small value of α. The reason why such feature was observed is that the small value of α corresponds to allowing the existence of permeable solid domain as described in Sect. 3.3.

For the crosscheck, the topology-optimized configurations were analyzed across the different α settings. Table 1 shows the crosscheck of the objective functional values for the different optimized configurations and α settings, and it is confirmed that the topology-optimized configuration for a certain condition is better than the others for its particular condition.

5 Conclusion

This paper discussed topology optimization formulated as a material distribution problem with respect to fluid and permeable solid, whereas the previous works dealing with topology optimization for the design of cooling systems have focused on fluid and impermeable solid. Through the numerical investigation, we found that the connectivity of fluid domain in the topology-optimized configuration is affected depending on the local values of inverse permeability α. That is, disconnected flow

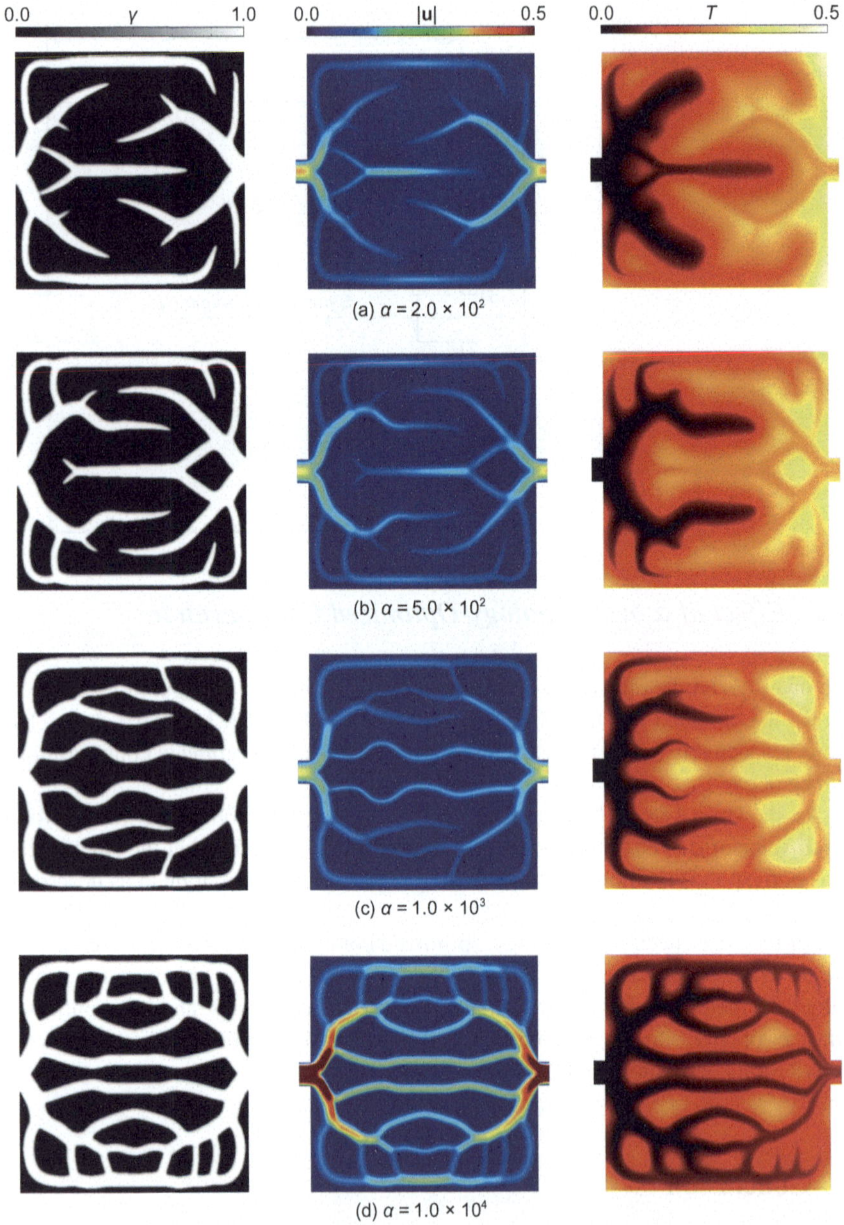

Fig. 3 Optimized configurations with the magnitude of velocity distributions and the temperature distributions for different α settings

Table 1 Crosscheck of J with respect to α in Fig. 3

Analysis α	Optimization α			
	2.0×10^2	5.0×10^2	1.0×10^3	1.0×10^4
2.0×10^2	**29.9**	29.2	27.7	25.0
5.0×10^2	25.6	**27.1**	26.3	24.6
1.0×10^3	20.9	24.6	**25.0**	24.5
1.0×10^4	4.83	10.6	17.6	**24.0**

channels were allowed as the topology-optimized configuration when setting a small value of α. This result indicates that the combination of disconnected flow channels and porous media holds a potential for boosting the heat transfer performance of a cooling system, as with the case of a previous work dealing with a flow battery design problem [30].

It remains a challenge for future research to perform the practical cases, e.g., three-dimensional problem and/or the use of practical porous model. Further insight into this aspect is left to future work.

Acknowledgments This work was partially supported by JSPS KAKENHI Grant Number 18K13674.

References

1. Alexandersen, J., Aage, N., Andreasen, C.S., Sigmund, O.: Topology optimisation for natural convection problems. Int. J. Numer. Meth. Fluids **76**(10), 699–721 (2014)
2. Alexandersen, J., Sigmund, O., Aage, N.: Large scale three-dimensional topology optimisation of heat sinks cooled by natural convection. Int. J. Heat. Mass. Trans. **100**, 876–891 (2016)
3. Bendsøe, M.P.: Optimal shape design as a material distribution problem. Struct. Optim. **1**(4), 193–202 (1989)
4. Bendsøe, M.P., Kikuchi, N.: Generating optimal topologies in structural design using a homogenization method. Comput. Methods Appl. Mech. Eng. **71**(2), 197–224 (1988)
5. Bendsøe, M.P., Sigmund, O.: Topology Optimization: Theory, Methods, and Applications. Springer, New York (2003)
6. Borrvall, T., Petersson, J.: Topology optimization of fluids in Stokes flow. Int. J. Numer. Meth. Fluids **41**(1), 77–107 (2003)
7. Challis, V.J., Guest, J.K.: Level set topology optimization of fluids in Stokes flow. Int. J. Numer. Methods. Eng. **79**(10), 1284–1308 (2009)
8. Coffin, P., Maute, K.: A level-set method for steady-state and transient natural convection problems. Struct. Multidisc. Optim. **53**(5), 1047–1067 (2016)
9. Deaton, J.D., Grandhi, R.V.: A survey of structural and multidisciplinary continuum topology optimization: post 2000. Struct. Multidisc. Opti. **49**(1), 1–38 (2014)
10. Deng, D., Tang, Y., Liang, D., He, H., Yang, S.: Flow boiling characteristics in porous heat sink with reentrant microchannels. Int. J. Heat. Mass. Trans. **70**, 463–477 (2014)
11. Dilgen, C.B., Dilgen, S.B., Fuhrman, D.R., Sigmund, O., Lazarov, B.S.: Topology optimization of turbulent flows. Comput. Methods Appl. Mech. Eng. **331**, 363–393 (2018)
12. Gersborg-Hansen, A., Sigmund, O., Haber, R.B.: Topology optimization of channel flow problems. Struct. Multidisc. Optim. **30**(3), 181–192 (2005)

13. Kawamoto, A., Matsumori, T., Yamasaki, S., Nomura, T., Kondoh, T., Nishiwaki, S.: Heaviside projection based topology optimization by a PDE-filtered scalar function. Struct. Multidisc. Optim. **44**(1), 19–24 (2011)
14. Koga, A.A., Lopes, E.C.C., Nova, H.F.V., de Lima, C.R., Silva, E.C.N.: Development of heat sink device by using topology optimization. Int. J. Heat Mass. Trans. **64**, 759–772 (2013)
15. Kreissl, S., Pingen, G., Maute, K.: Topology optimization for unsteady flow. Int. J. Numer. Meth. Eng. **87**(13), 1229–1253 (2011)
16. Kubo, S., Yaji, K., Yamada, T., Izui, K., Nishiwaki, S.: A level set-based topology optimization method for optimal manifold designs with flow uniformity in plate-type microchannel reactors. Struct. Multidisc. Optim. **55**(4), 1311–1327 (2017)
17. Lazarov, B.S., Sigmund, O.: Filters in topology optimization based on Helmholtz-type differential equations. Int. J. Numer. Meth. Eng. **86**(6), 765–781 (2011)
18. Lin, S., Zhao, L., Guest, J.K., Weihs, T.P., Liu, Z.: Topology optimization of fixed-geometry fluid diodes. J. Mech. Des. **137**(8), 081, 402 (2015)
19. Matsumori, T., Kondoh, T., Kawamoto, A., Nomura, T.: Topology optimization for fluid-thermal interaction problems under constant input power. Struct. Multidisc. Optim. **47**(4), 571–581 (2013)
20. Sato, Y., Yaji, K., Izui, K., Yamada, T., Nishiwaki, S.: Topology optimization of a no-moving-part valve incorporating Pareto frontier exploration. Struct. Multidisc. Opti. **56**(4), 839–851 (2017)
21. Sato, Y., Yaji, K., Izui, K., Yamada, T., Nishiwaki, S.: An optimum design method for a thermal-fluid device incorporating multiobjective topology optimization with an adaptive weighting scheme. J. Mech. Design. **140**(3), 031, 402 (2018)
22. Sigmund, O., Maute, K.: Topology optimization approaches. Struct. Multidisc. Optim. **48**(6), 1031–1055 (2013)
23. Singh, R., Akbarzadeh, A., Mochizuki, M.: Sintered porous heat sink for cooling of high-powered microprocessors for server applications. Int. J. Heat Mass. Trans. **52**(9–10), 2289–2299 (2009)
24. Tezduyar, T.E., Mittal, S., Ray, S., Shih, R.: Incompressible flow computations with stabilized bilinear and linear equal-order-interpolation velocity-pressure elements. Comput. Methods Appl. Mech. Eng. **95**(2), 221–242 (1992)
25. Tomadakis, M.M., Robertson, T.J.: Viscous permeability of random fiber structures: comparison of electrical and diffusional estimates with experimental and analytical results. J. Compos. Mater. **39**(2), 163–188 (2005)
26. Wan, Z., Guo, G., Su, K., Tu, Z., Liu, W.: Experimental analysis of flow and heat transfer in a miniature porous heat sink for high heat flux application. Int. J. Heat Mass Trans. **55**(15–16), 4437–4441 (2012)
27. Yaji, K., Ogino, M., Chen, C., Fujita, K.: Large-scale topology optimization incorporating local-in-time adjoint-based method for unsteady thermal-fluid problem. Struct. Multidisc. Opti. **58**(2), 817–822 (2018)
28. Yaji, K., Yamada, T., Kubo, S., Izui, K., Nishiwaki, S.: A topology optimization method for a coupled thermal-fluid problem using level set boundary expressions. Int. J. Heat Mass Trans. **81**, 878–888 (2015)
29. Yaji, K., Yamada, T., Yoshino, M., Matsumoto, T., Izui, K., Nishiwaki, S.: Topology optimization in thermal-fluid flow using the lattice Boltzmann method. J. Comput. Phys. **307**, 355–377 (2016)
30. Yaji, K., Yamasaki, S., Tsushima, S., Suzuki, T., Fujita, K.: Topology optimization for the design of flow fields in a redox flow battery. Struct. Multidisc. Opti. **57**(2), 535–546 (2018)
31. Zhao, X., Zhou, M., Sigmund, O., Andreasen, C.S.: A "poor man's approach" to topology optimization of cooling channels based on a Darcy flow model. Int. J. Heat Mass Trans. **116**, 1108–1123 (2018)

Earthquakes and Tsunamis

Quasi-static Simulation Method of Earthquake Cycles Based on Viscoelastic Finite Element Modeling

Ryoichiro Agata, Takane Hori, Sylvain D. Barbot, Mamoru Hyodo, and Tsuyoshi Ichimura

Abstract Earthquake cycle simulations are important for studying earthquake generation processes and physics-based earthquake damage estimations. Earthquake cycle simulation methods typically assume a frictional constitutive relation on a known fault plane in a solid continuum and calculate earthquake evolution as spontaneous fault slip. To carry out such simulations, the boundary integral equation method, based on an elastic half-space, is widely used. In this approach, stress change around the fault plane due to crustal deformation can be computed analytically, but physical properties such as three-dimensional heterogeneous structure and viscoelastic deformation in mantle are generally not taken into account. Here, we incorporate such complex physical properties in the earthquake cycle simulation based on finite element modeling, using state-of-the-art techniques in computational science. We apply the proposed method to a fundamental problem of earthquake cycle generation and obtain results consistent with past studies.

R. Agata (✉) · T. Hori
Japan Agency for Marine-Earth Science and Technology, Kanagawa, Japan
e-mail: agatar@jamstec.go.jp

T. Hori
e-mail: horit@jamstec.go.jp

M. Hyodo
Japan Agency for Marine-Earth Science and Technology, (Now at Japan Meteorological Agency), Kanagawa, Japan
e-mail: hyodo@jamstec.go.jp

S. D. Barbot
Nanyang Technological University, Singapore (Now at University of Southern California), Nanyang, Singapore
e-mail: sylbar.vainbot@gmail.com

T. Ichimura
The University of Tokyo, (Now also at RIKEN), Tokyo, Japan
e-mail: ichimura@eri.u-tokyo.ac.jp

H. Itou et al. (eds.), *Mathematical Analysis of Continuum Mechanics and Industrial Applications III*, Mathematics for Industry 34,
https://doi.org/10.1007/978-981-15-6062-0_11

1 Introduction

Earthquake cycle simulation methods have been widely applied to explaining earthquake generation processes. They are also expected to play an important role in disaster mitigation, such as in generation of possible earthquake scenarios. Typical earthquake cycle simulation methods assume that a fault plane is embedded in a known location within a solid continuum. Slip evolution on the fault plane due to the driving force from the relative motion of two sides of the plane is assumed to be subjected to a frictional constitutive relation. By solving the equations for the frictional constitutive relation and equilibrium of forces in the media, earthquake evolution is calculated as spontaneous fault slip. The earthquake cycle simulation methods used in past studies have mostly been based on homogeneous elastic half-space, in which stress interaction around the fault plane is analytically obtained [16]. This allows for calculation of spontaneous fault slip with moderate computation cost based on the boundary integral equation method (BIEM) (e.g., [18]). On the other hand, subduction zones, where fault regions of magnitude-9-class earthquakes are located, are of fully three-dimensional (3D) heterogeneous structure. It is also known that viscoelastic flow in the upper mantle, which is located within approximately 30–200 km depth, is not negligible in a long-term deformation lasting for more than a few years. The BIEM approach with homogeneous elastic half-space is generally not capable of considering these properties. Calculating stress interaction around the fault plane based on numerical simulation methods such as the finite element (FE) method is a possible approach to incorporate such properties. Since FE simulations of earthquake cycles are known to be associated with significantly larger computation cost than in BIEM, application has been limited to simple two-dimensional (2D) problems (e.g., [9]). However, recent developments toward faster and more scalable FE solvers suitable for supercomputers [7, 8] are expected to overcome the issue of computation cost.

For this purpose, we have worked on incorporating FE modeling of viscoelastic deformation in earthquake cycle simulation with the aid of recent developments in computational science [7, 8]. We successfully applied the developed method to a simulation of deformation after the 2011 Tohoku-Oki earthquake, which is likely to have resulted from a combination of nonlinear viscoelastic relaxation and afterslip (transient fault slip that occurs following an earthquake) [1], as afterslip evolution can be interpreted as part of the earthquake cycle. This paper aims to explain the formulation and technical aspect of the proposed method in detail and present a demonstrative application example. In Sect. 2, we present the methodology. Application to a fundamental problem setting will be presented in Sect. 3. Section 4 provides concluding remarks.

2 Methods

Figure 1a is a 2D schematic view of a subduction zone, which is a typical target of earthquake cycle simulations. A subduction zone consists of two plates, one of which is subducting beneath another. A relatively shallow portion of the interface between the plates is thought to host earthquakes due to certain fault friction properties. The upper layer, in which brittle rocks are located, is called the crust. The deeper portion is called the mantle, which shows more ductile behavior due to high temperature. Our strategy for modeling earthquake cycles in a subduction zone is to assume viscoelastic material in a limited domain and an embedded fault plane within the material, where we apply the rate- and state-dependent friction law [4, 19], one of the most widely used friction models in such simulations, based on a widely accepted assumption to use the steady-state subduction as the reference state, in which the deformation is defined to be zero [20]. Then we only have to consider the deformation due to fault slips as perturbation from the reference state, which enables us to solve the problem using the FE method based on small deformation theory (Fig. 1b). In this section, we describe this modeling strategy in detail.

2.1 *Finite Element Formulation of Viscoelastic Deformation*

We assume that rocks that compose the material surrounding fault planes obey a viscoelastic constitutive relationship. Assuming a small deformation problem, force equilibrium equation and infinitesimal strain are written as (hereafter we use the Einstein summation convention)

$$\sigma_{ij,j} + f_i = 0, \tag{1}$$

$$\varepsilon_{ij} = \frac{1}{2}(u_{i,j} + u_{j,i}), \tag{2}$$

where σ_{ij}, f_i, ε_{ij}, and u_i are total stress tensor, outer force vector, strain tensor, and displacement vector, respectively. Note that strain, stress, and displacement are defined as the perturbation from the reference state, which is discussed later in detail. Since we assume that long-term deformation of rock is quasi-static, no acceleration term is included. The theory on which our assumption of small deformation is based

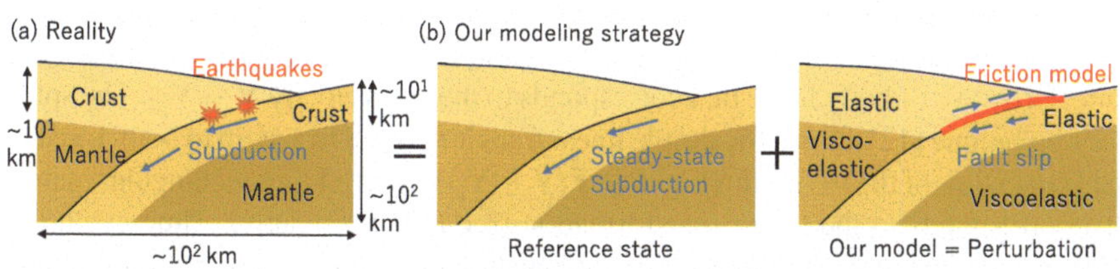

Fig. 1 **a** 2D schematic view of a subduction. **b** Overview of our modeling strategy

will be described later. For the constitutive relation of viscoelastic material, we assume the Maxwell model, as

$$\varepsilon_{ij} = \varepsilon^{\mathrm{e}}_{ij} + \varepsilon^{\mathrm{i}}_{ij}, \tag{3}$$

$$\sigma_{ij} = C_{ijkl}\varepsilon^{\mathrm{e}}_{kl}, \tag{4}$$

$$\dot{\varepsilon}^{\mathrm{i}}_{ij} = \frac{1}{2\eta}\sigma'_{ij}, \tag{5}$$

where $\varepsilon^{\mathrm{e}}_{ij}, \varepsilon^{\mathrm{i}}_{ij}, C_{ijkl}, \eta$ are elastic strain tensor, viscous strain tensor, elastic tensor, and viscosity. σ'_{ij} is deviatoric stress, calculated as $\sigma_{ij} - \frac{1}{3}\delta_{ij}\sigma_{kk}$. In this paper, for simplicity, we only discuss viscosity which is temporally constant, i.e., linear viscoelasticity. Substituting (3) into the time derivative of (1) yields

$$(C_{ijkl}\dot{\varepsilon}_{kl})_{,j} = -\dot{f}_i + (C_{ijkl}\dot{\varepsilon}^{\mathrm{i}}_{kl})_{,j}. \tag{6}$$

By discretizing the weak form of this equation with proper boundary conditions based on the FE method, we obtain the linear equation

$$\mathbf{K}\dot{\mathbf{u}} = \dot{\mathbf{f}} + \dot{\mathbf{h}}(\dot{\boldsymbol{\varepsilon}}^{\mathrm{i}}), \tag{7}$$

where $\mathbf{K}, \mathbf{u}, \mathbf{f}$, and $\mathbf{h}$ are global stiffness matrix, displacement vector, outer force vector, and vector corresponding to viscoelastic stress relaxation. Dirichlet boundary conditions are imposed at the sides and bottom of the targeted domain. Specifically, displacements in the normal direction at points on both the side and bottom planes are fixed to zero, while there are no imposed values in other directions. $\dot{\mathbf{h}}(\dot{\boldsymbol{\varepsilon}}^{\mathrm{i}})$ denotes that $\dot{\mathbf{h}}$ is a function of a vector of the inelastic strain tensor rate.

Viscoelastic deformation is driven by slip velocity on the fault plane. Here, we define "slip" as the gap of displacements on points at two sides of the fault plane. Hereafter, we use V_k to denote slip velocity on the FEM node k on the fault, assuming the direction of slip is known. $\mathbf{V}$ is a slip velocity vector comprising V_k. Here, we consider the fault source based on the back-slip model [20], which assumes the steady-state subduction motion as the reference state, i.e., a state with no strain, although possibly associated with high stress. We consider only the perturbation from this reference state: $\mathbf{V} - \mathbf{V}_{\mathrm{pl}}$, the perturbation of $\mathbf{V}$ from $\mathbf{V}_{\mathrm{pl}}$, is the source of deformation, where $\mathbf{V}_{\mathrm{pl}}$ is velocity of the steady-state subduction. No deformation occurs when $\mathbf{V} = \mathbf{V}_{\mathrm{pl}}$. This assumption allows for solving the problem based on small deformation theory, meaning the governing equation becomes considerably simplified because of linearization and use of single reference coordinate frame. $\dot{\mathbf{f}}$ is time derivative of body force that corresponds to the slip velocity $\mathbf{V} - \mathbf{V}_{\mathrm{pl}}$. The split-node technique [12] allows for simple linear transformation from slip to body force in the framework of the FE method, so $\dot{\mathbf{f}} = \mathbf{F}(\mathbf{V} - \mathbf{V}_{\mathrm{pl}})$, where $\mathbf{F}$ is a rectangular matrix that corresponds to the linear transformation (see [12] for details). Time evolution of $\mathbf{V}$ is calculated based on the frictional constitutive relation explained in the next subsection.

When 3D heterogeneous structure is incorporated, the linear equation (7) typically has $10^7 - 10^9$ degrees of freedom (DOF). To solve such a large equation, we use the fast, scalable, and memory-efficient solver that runs on a massively parallel computation environment proposed by [7, 8] and extended by [2]. Because direct solvers consume too much memory for a large-scale problem, the conjugate gradient (CG) method, a widely used iterative solver, is used. The main computational cost of the CG method is a repeat of the computing matrix–vector product, for example, $\mathbf{Kv}$, where $\mathbf{v}$ is an arbitrary vector. To maintain memory efficiency, the matrix–vector product, with $\mathbf{K}$ in the CG algorithm, is calculated without storing the entire $\mathbf{K}$ matrix in the memory based on the element-by-element (EBE) method [21]. In the solver, all the matrix operations described in this section, such as matrix–vector products and vector–vector products, are computed in parallel using MPI and OpenMP based on the domain decomposition using Metis [10], which enables good load balancing for all MPI processes.

2.2 The Rate- and State-Dependent Friction Law

We incorporate the rate- and state-dependent friction law, which is derived from laboratory rock experiments [4, 19]. This friction law well explains the experimental data obtained with the slip rate below 10^{-3} m/s, assuming that friction coefficient depends on both slip velocity (rate) and a state variable. Note that this friction law does not help detailed discussions on earthquake rupture, in which the slip rate is typically in the order of 1 m/s. The state variable is analogous to "strength as a threshold" [14], which introduces both the weakening and strength-recovery (healing) processes of frictional strength. We assume the equilibrium equation in shear stress on the fault plane, which is described as

$$\frac{\mathrm{d}\tau_k}{\mathrm{d}t} = F_k(\mathbf{V} - \mathbf{V}_{\mathrm{pl}}, \dot{\boldsymbol{\varepsilon}}^{\mathrm{i}}) - \gamma \frac{\mathrm{d}V_k}{\mathrm{d}t}, \tag{8}$$

where τ_k is shear stress on the FEM node k on the fault, F_k is a function that maps $\mathbf{V} - \mathbf{V}_{\mathrm{pl}}$ and $\boldsymbol{\varepsilon}^{\mathrm{i}}$ to shear stress change on the node k, and $\boldsymbol{\varepsilon}^{\mathrm{i}}$ is inelastic strain in the targeted 3D body. The acceleration term, which is significant only when slip rate is large, is not included in the equation. Instead, the second term in the right-hand side introduces the effect of the seismic radiation damping and represents energy outflow as seismic wave [18]. $\gamma = \mu/2c$, where μ is the rigidity and c is the shear wave velocity. In many studies, simulations with elastic homogeneous half-space based on BIEM were carried out, where $\dot{\boldsymbol{\varepsilon}}^{\mathrm{i}} = 0$. Then F_k can be written as a convolutional operation $K_{kl}(V_l - V_{\mathrm{pl}})$, where K_{kl} is stress change at the k-th node due to unit displacement at the l-th node. K_{kl} in elastic homogeneous half-space can be computed based on analytical expressions (e.g., [16]), which leads to relatively small computation cost associated with BIEM. In this study, we evaluate F_k directly from $\dot{\mathbf{u}}$ computed by using the FEM, in which F_k can be a function of both $\mathbf{V}$ and $\dot{\boldsymbol{\varepsilon}}^{\mathrm{i}}$, and arbitrary geometry and material heterogeneity can be considered. It should be noted that a BIEM approach that can incorporate inelastic strain in elastic homogeneous half-space was proposed recently [3, 11]. The equations for the rate- and state-dependent friction law used to model frictional behavior on the plate interface are

$$V_k = V_* \exp\left(\frac{\tau_k - (\tau_{s*k} + \Delta\tau_{sk})}{A_k}\right), \tag{9}$$

$$\frac{d\Delta\tau_{sk}}{dt} = \frac{B_k}{L_k/V_*}\exp\left(-\frac{\Delta\tau_{sk}}{B_k}\right) - \frac{B_k V_k}{L_k}. \tag{10}$$

(9) represents a fault constitutive law that determines V_k for a given τ_k and a value of $\tau_{sk}(=\tau_{s*k} + \Delta\tau_{sk})$, where $\Delta\tau_{sk}$ is a state variable and we set $V_* = V_{\mathrm{pl}}$. τ_{s*k} is the steady-state strength with $V_k = V_{\mathrm{pl}}$, i.e., $\frac{d\Delta\tau_{sk}}{dt} = 0$ and $V_k = V_{\mathrm{pl}}$ require $\Delta\tau_{sk} = 0$, which results in $V_k = V_{\mathrm{pl}} = V_{\mathrm{pl}} \exp\left(\frac{\tau_k - \tau_{s*k}}{A_k}\right)$. Therefore, $\tau_{s*k} = \tau|_{V_k = V_{\mathrm{pl}}}$ is the stress when $V_k = V_{\mathrm{pl}}$ holds. (10) is an aging law [19]. The frictional parameter B controls strength recovery, while L controls slip weakening.

Deleting V_k from (8) to (10) using (9), we obtain four ordinary differential equations: (5), (7), (8), and (10). Internal deformation in the solid media is calculated in (5) and (7) based on FE modeling. Equations (8) and (10) provide the conditions for time evolution of the variables on the embedded fault plane, which interact with the internal deformation. We perform time integration using an adaptive time-step fifth-order Runge–Kutta algorithm [17].

3 Application

We perform earthquake cycle simulations in a fundamental problem setting. The purpose of these simulations is to show the applicability of the proposed method and the effect of introducing viscoelasticity.

3.1 Problem Setting

We consider a horizontal planar fault embedded in a viscoelastic material. The fault plane has 304 km width and length, in which friction parameters are set following [15]. Distribution of frictional parameter $A - B$ is shown in Fig. 2a, where $V_* = V_{\mathrm{pl}} = 10^{-9}$ m/s with a fixed slip direction and $L = 0.17$ m uniformly. The friction parameters are all constant. We consider $A - B$ instead of A and B because the sign of $A - B$ determines stability of slip. Within such a setting, unstable fast slips that correspond to earthquakes are expected to occur at the central part of the fault plane where the value of $A - B$ is negative [19]. Stable slow slips are expected to

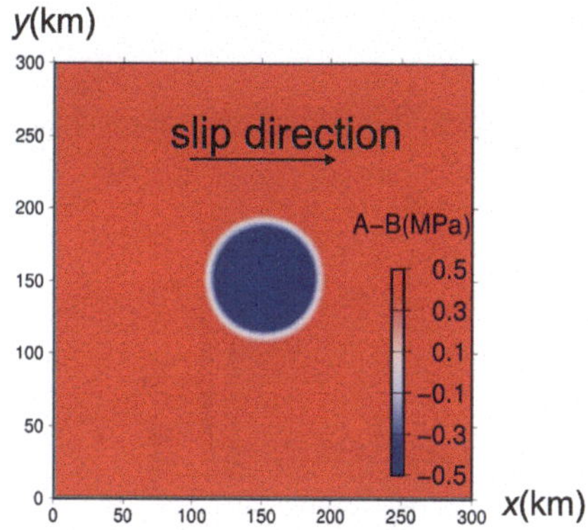

(a)

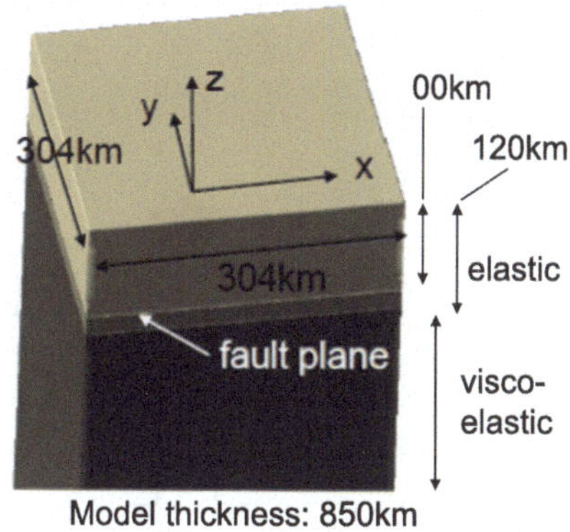

(b)

Fig. 2 Simulation settings. **a** Distribution of frictional parameter $A - B$. We plot this quantity instead of A and B because the sign of $A - B$ determines stability of slip. Within such a setting, unstable slip is expected to occur at the center of the fault plane, where $A - B < 0$. **b** FE model for the target region. The fault is embedded at the depth of 100 km in a two-layered material, whose first layer is elastic and 120 km thick, and whose second layer is viscoelastic

occur elsewhere. The fault is embedded at the depth of 100 km in a two-layered material, whose first layer is elastic and 120 km thick and the second viscoelastic. The elastic layer corresponds to crust which hosts brittle failure, and the viscoelastic layer to mantle in which ductile failure takes place. Although the proposed method can incorporate elastic heterogeneity, elastic parameters here are homogeneously set for simplicity as $\mu = 30\,\text{GPa}$ and $\nu = 0.25$, where μ is rigidity and ν is Poisson's ratio. Figure 2b shows the FE model that we constructed for the target problem. The fault plane is discretized by FE nodes in 0.5 km resolution. This setting results in an FE model of 33, 614, 073 DOF and 8, 181, 704 tetrahedral elements.

The initial value of variables in (8), (9), and (10) are as follows: $V = 0.9V_{\text{pl}}$ and $\Delta\tau_s = -B\ln(0.9V_{\text{pl}}/V_{\text{pl}})$ [6]. The initial τ is computed accordingly based on (9). We compare the simulations of different viscosity η in the second layer: one with $\eta = \infty$ (hereafter called elastic model) and the other $\eta = 1 \times 10^{19}\,\text{Pa s}$ (hereafter called viscoelastic model). Thus, we examine the effect of introducing viscoelasticity on earthquake occurrence. To perform the calculation, we use the K computer at the RIKEN Center for Computational Science [13], each computer node of which has one CPU (Fujitsu SPARC64 VIIIfx 8 core 2.0 GHz) and 16 GB of memory.

3.2 Result

Figure 3 shows the time evolution of slip velocity on the fault plane calculated in the elastic model. As expected based on the parameter setting, repeated earthquake occurrence is observed at the center of the fault plane: fast slip is indicated by the red

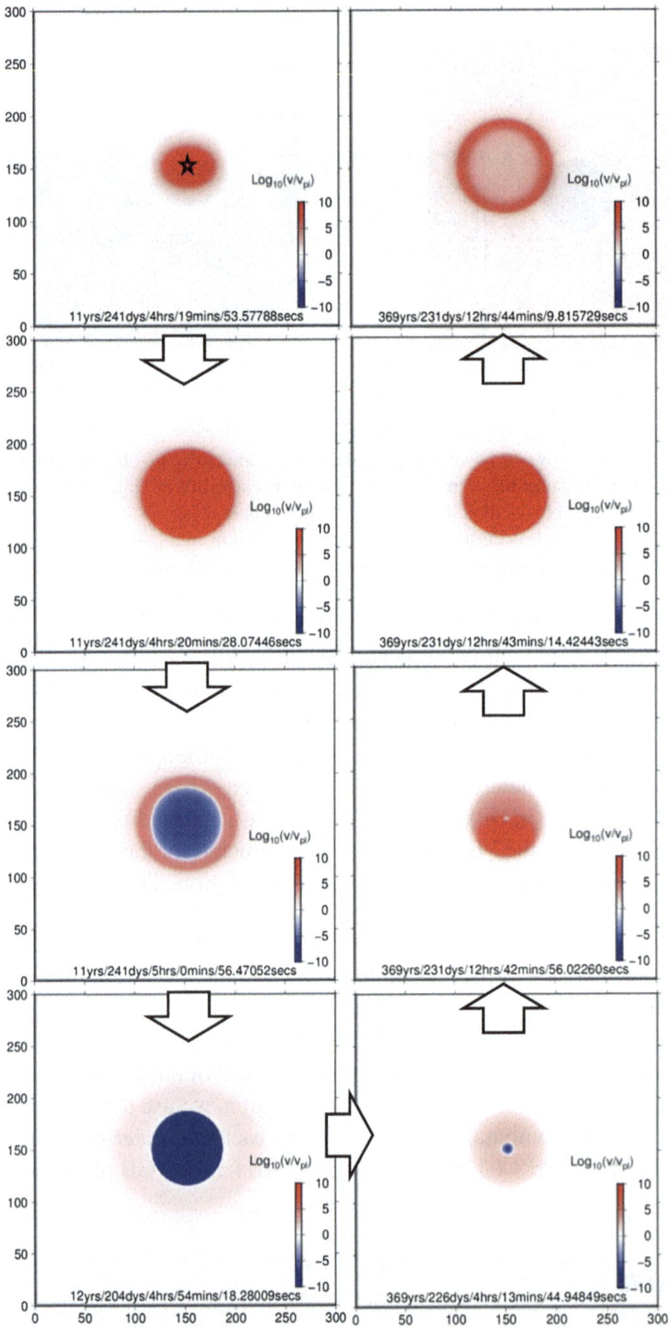

Fig. 3 The time evolution of slip velocity on the fault plane calculated in the elastic model. Fast slip indicated by the red color corresponds to earthquake, and slip with tiny velocity indicated by the blue color corresponds to coupling

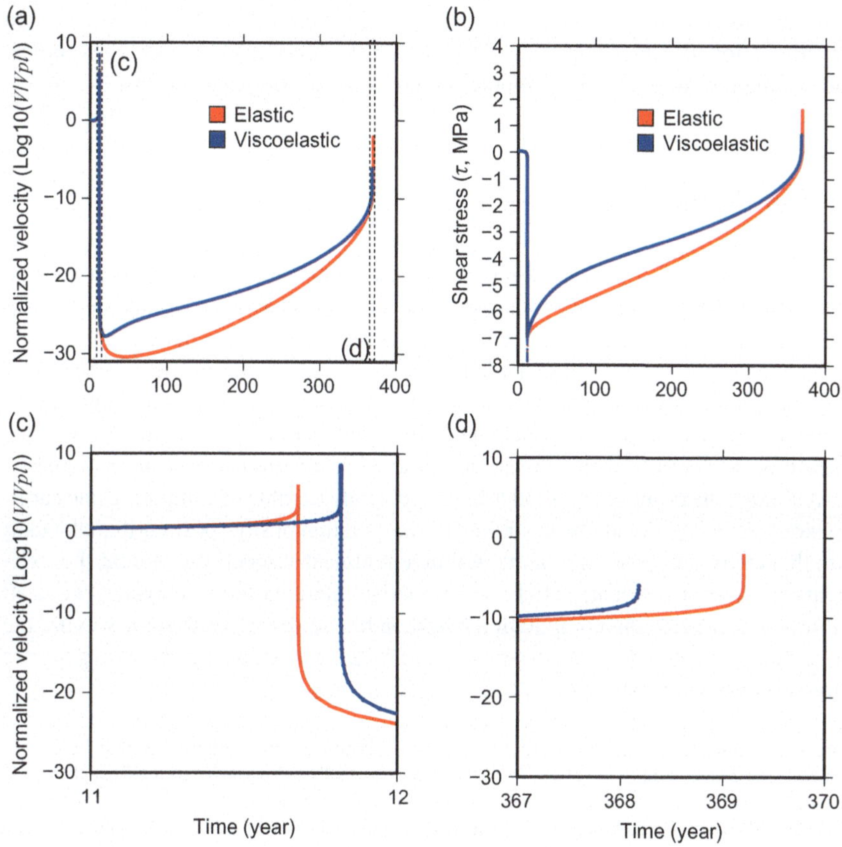

Fig. 4 The time history of the logarithm of normalized slip velocity **a** and stress **b** in the central point of the fault plane, denoted by the star in Fig. 3. **c** and **d** are the close-up views of the slip rate history within the dashed squares denoted in (**a**)

color and strong coupling, in which relative velocity of the two sides of the fault plane is almost zero, by the blue color. This result is consistent with previous knowledge about such simulations based on the rate- and state-dependent friction law with an elastic material.

Figure 4 shows the time history of slip velocity and stress in the central point of the fault plane. Here, we show the results with both the elastic and the viscoelastic models. The first occurrence of the earthquake in the viscoelastic model is delayed, probably because viscoelastic relaxation slows down the stress accumulation on the fault plane. After the first earthquake, viscoelastic relaxation accelerates the stress accumulation on the fault plane (Fig. 4b), which results in the earlier occurrence of the second earthquake in the viscoelastic model (Fig. 4d). As a result, recurrence time of earthquakes in the viscoelastic model becomes shorter than in the elastic one. In the simulation, the linear equation (7) is solved 10^5 times in each case. To complete the calculation, we used 385 computer nodes of the K computer for 24 h.

4 Concluding Remarks

This paper described the development of an earthquake cycle simulation method based on 3D FE modeling. Combining a viscoelastic and a rate- and state-dependent friction model, we obtained a large linear equation with a sparse matrix. We applied a fast, memory-efficient, and scalable FE solver that allows the equation to be solved fast enough for earthquake cycle simulations. The application example within a fundamental problem setting showed that the proposed method could simulate one cycle of earthquakes. Viscoelastic relaxation shortened the recurrence time of earthquakes.

Although the proposed method is tentatively only capable of computing one earthquake cycle, it has already been applied to computation of slow post-earthquake fault slip that evolves over several years within a nonlinear viscoelastic mantle (i.e., viscosity is stress-dependent) [1]. To accommodate more earthquake cycles, we need to further accelerate the computation based on the newest algorithm for solving the target linear equation [5]. We also plan to improve the meshing algorithm to generate more efficient FE mesh.

Acknowledgments We thank the reviewer for his comments and suggestions that helped us improve the manuscript. This study was supported by JSPS Fellowship (26-8867) and Post K computer project (Priority issue 3: Development of Integrated Simulation Systems for Hazard and Disaster Induced by Earthquake and Tsunami). We obtained the results using the K computer at the RIKEN Center for Computational Science (Proposal number hp160221 and hp170249).

References

1. Agata, R., Barbot, S.D., Fujita, K., Hyodo, M., Iinuma, T., Nakata, R., Ichimura, T., Hori, T.: Rapid mantle flow with power-law creep explains deformation after the 2011 tohoku megaquake, Nature Communications, **10**(1), 1385 (2018)
2. Agata, R., Ichimura, T., Hori, T., Hirahara, K., Hashimoto, C., Hori, M.: An adjoint-based simultaneous estimation method of the asthenosphere's viscosity and afterslip using a fast and scalable finite-element adjoint solver. Geophys. J. Int. **213**(1), 461–474 (2018)
3. Barbot, S.: Asthenosphere flow modulated by megathrust earthquake cycles. Geophys. Res. Lett. **45**(12) (2018)
4. Dieterich J.H.: Modeling of rock friction: 1. experimental results and constitutive equations. J. Geophys. Res.: Solid Earth **84**(B5), 2161–2168 (1979)
5. Fujita, K., Ichimura, T., Koyama, K., Inoue, H., Hori, M., Maddegedara, L.: Fast and scalable low-order implicit unstructured finite-element solver for earth's crust deformation problem. In: Proceedings of the Platform for Advanced Scientific Computing Conference, p. 11. ACM (2017)

6. Hyodo, M., Hori, T.: Re-examination of possible great interplate earthquake scenarios in the Nankai Trough, southwest Japan, based on recent findings and numerical simulations. Tectonophysics **600**, 175–186 (2013)
7. Ichimura, T., Agata, R., Hori, T., Hirahara, K., Hashimoto, C., Hori, M., Fukahata, Y.: An elastic/viscoelastic finite element analysis method for crustal deformation using a 3-D island-scale high-fidelity model. Geophys. J. Int. **206**(1), 114–129 (2016)
8. Ichimura, T., Fujita, K., Tanaka, S., Hori, M., Lalith, M., Shizawa, Y., Kobayashi, H.: Physics-based urban earthquake simulation enhanced by 10.7 BlnDOF x 30 K time-step unstructured FE non-linear seismic wave simulation. In: SC14: International Conference for High Performance Computing, Networking, Storage and Analysis, pp. 15–26 (2014)
9. Kaneko, Y., Ampuero, J.-P., Lapusta, N.: Spectral-element simulations of long-term fault slip: Effect of low-rigidity layers on earthquake-cycle dynamics. J. Geophys. Res.: Solid Earth **116**(B10) (2011)
10. Karypis, G., Kumar, V.: A fast and high quality multilevel scheme for partitioning irregular graphs. SIAM J. Sci. Comput. **20**(1), 359–392 (1998)
11. Lambert, V., Barbot, S.: Contribution of viscoelastic flow in earthquake cycles within the lithosphere-asthenosphere system. Geophys. Res. Lett. **43**(19), 10,142–10,154 (2016)
12. Melosh, H.J., Raefsky, A.: A simple and efficient method for introducing faults into finite element computations. Bull. Seismol. Soc. Am. **71**(5), 1391–1400 (1981)
13. Miyazaki, H., Kusano, Y., Shinjou, N., Shoji, F., Yokokawa, M., Watanabe, T.: Overview of the K computer system. Fujitsu Sci. Tech. J. **48**(3), 255–265 (2012)
14. Nakatani, M.: Conceptual and physical clarification of rate and state friction: frictional sliding as a thermally activated rheology. **106**, 1 (2001)
15. Noda, H., Hori, T.: Under what circumstances does a seismogenic patch produce aseismic transients in the later interseismic period? Geophys. Res. Lett. **41**(21), 7477–7484 (2014)
16. Okada, Y.: Internal deformation due to shear and tensile faults in a half-space. Bull. Seismol. Soc. Am. **82**(2), 1018–1040 (1992)
17. Press, W.H., Teukolsky, S.A., Vetterling, W.T., Flannery, B.P.: Numerical Recipes in C (2nd ed.): The Art of Scientific Computing. Cambridge University Press, New York (1992)
18. Rice, J.R.: Spatio-temporal complexity of slip on a fault. J. Geophys. Res.: Solid Earth **98**(B6), 9885–9907 (1993)
19. Ruina, A.: Slip instability and state variable friction laws. J. Geophys. Res.: Solid Earth **88**(B12), 10359–10370 (1983)
20. Savage, J.C.: A dislocation model of strain accumulation and release at a subduction zone. J. Geophys. Res.: Solid Earth **88**(B6), 4984–4996 (1983)
21. Winget, J.M., Hughes, T.J.R.: Solution algorithms for nonlinear transient heat conduction analysis employing element-by-element iterative strategies. Comput. Methods Appl. Mech. Eng. **52**(1–3), 711–815 (1985)

Friction Versus Damage: Dynamic Self-similar Crack Growth Revisited

Shiro Hirano

Abstract Seismological observational studies have revealed that earthquakes exhibit dynamic self-similar crack growth constituting 50–90% of the shear wave velocity. Remarkably, the peak slip velocity defined on the crack surface is scale-invariant, even from $M1$ to $M9$ earthquakes. However, a classical self-similar crack model with a singularity does not satisfy all the observed properties above. In this chapter, we review these discrepancies and introduce friction and damage models to solve them, which have been proposed in several numerical studies. We show that velocity-dependent friction can fulfill some requirements of the observations, while slip- or time-dependent friction cannot. We finally discuss the theoretical equivalence of friction and damage model for a self-similar crack in terms of energetics, which has previously only been implied by numerical studies.

1 Introduction: Applications and Limitations of a Classical Crack Model for Earthquake Mechanics

Earthquakes involve dynamic rupture propagation with speed of 2–3 km/s along faults, which are shear cracks embedded in the Earth's crust. To model this faulting process mathematically, we need to define the amount of slip on faults as a function of space and time and the magnitude of earthquakes. Let $u(x,t) \in \mathbb{R}^3$ be displacement of the medium defined at position $x \in \mathbb{R}^3 \backslash \Gamma$, where t is time since the initiation of dynamic rupture and Γ is a fault surface. Then, we define slip $D(x,t)$ and slip velocity $V(x,t)$ on Γ as

$$D(x,t) := \lim_{\varepsilon \downarrow 0} u\left(x + \varepsilon \nu^{\Gamma}, t\right) - \lim_{\varepsilon \uparrow 0} u\left(x + \varepsilon \nu^{\Gamma}, t\right), \qquad (1)$$

$$V(x,t) := \partial_t D(x,t), \qquad (2)$$

S. Hirano (✉)
College of Science and Engineering, Ritsumeikan University, 1-1-1, Noji Higashi, Kusatsu, Shiga 525-8577, Japan
e-mail: s-hrn@fc.ritsumei.ac.jp

H. Itou et al. (eds.), *Mathematical Analysis of Continuum Mechanics and Industrial Applications III*, Mathematics for Industry 34,
https://doi.org/10.1007/978-981-15-6062-0_12

respectively, where ν^Γ is a unit normal vector to Γ and $D = V = 0$ is assumed for $t \leq 0$. Earthquake size is quantified by the seismic moment, M_0, originally defined as [1]

$$M_0 := \mu \int_\Gamma |D(x, T)| \, \mathrm{d}x,$$

where μ is the rigidity, and T is the duration of the earthquake, i.e., the amount of time required to generate the final state of the earthquake process. By extending this definition, the time-dependent seismic moment function $M_0(t)$ can be defined as

$$M_0(t) := \mu \int_\Gamma |D(x, t)| \, \mathrm{d}x. \tag{3}$$

The static M_0 is proportional to the minimum value of elastic strain energy released by the earthquake if the on-fault traction change is spatially constant; see Hirano [12] and references therein for a precise review and practical problems with this relationship. Additionally, the (seismic moment) magnitude $M_w := \frac{2}{3}\left(\log_{10} M_0 - 9.1\right)$ is widely employed.

The duration, T, can reach $\sim 10^2$–10^3 s for $M9$ earthquakes, which are the largest events on the earth, while T is at most ~ 1 and $\sim 10^{-6}$ s for $M4$ (medium-sized in nature) and $M - 8$ (smallest events observed in the laboratory) earthquakes, respectively. Hence, earthquake duration varies 10^9 fold from the smallest to the largest. A question then arises: are the mechanisms of small, medium, and large earthquakes fundamentally different? Alternatively, do they obey a universal law? In terms of kinematics, the empirical relationship among them strongly suggests the existence of a governing law. Over almost the entire range of natural and laboratory earthquakes, the seismic moment, M_0, and corner frequency (or cutoff frequency in engineering terminology), $f_c \sim 1/T$, show the following relationship: $M_0 \propto f_c^{-3}$ [24, 27]. In other words, T must scale as $M_0 \propto T^3$ [15]. Moreover, this relationship is observed not only in the final state but also in a temporal state as

$$M_0(t) \propto t^3 \tag{4}$$

at least for the Parkfield area in California [23] and Japan [20], which means that the system is self-similar [5]. If the rupture propagation velocity is almost constant and given as v_r, the current crack radius is written as $r = v_r t$, and the scaling relationship (4) suggests that the macroscopic energy release rate, $G := \mathrm{d}M_0/\mathrm{d}A$, of the system satisfies

$$G \propto F(v_r)\frac{\mathrm{d}A^{2/3}}{\mathrm{d}A} \propto F(v_r) r = v_r F(v_r) t, \tag{5}$$

where $A(\propto r^2)$ is the current area of the ruptured region. This proportionality (5) and the function F are simply derived from modeling the self-similar expansion of circular and elliptic singular cracks (Sect. 6.9 of Broberg [6]). In fact, a zeroth-order approximation of the earthquake faulting process has been provided by such a singular self-similar crack model.

However, a precise comparison of observations and the singular crack model reveals some discrepancies. In a traditional framework of linear elastic fracture mechanics, the energy release rate, G, must be balanced with the surface energy of the material, 2γ. If γ is a material constant, the relationship (5) never balances with 2γ unless the rupture propagation velocity asymptotes toward the Rayleigh wave speed, c_R, for mode-II ruptures or the shear wave speed, c_S, for mode-III ruptures, because $\lim_{v_r \to c_R, c_S} F(v_r) = 0$ [6]. On the contrary, the typical rupture propagation velocity of earthquakes is 50–90 % of the Rayleigh or shear wave speed of Earth's crust (e.g., [11, 18]). Thus, both the release and dissipation rate of energy must increase during dynamic rupture growth.

Another discrepancy is as follows. The singular crack model means that the slip velocity diverges in the vicinity of the rupture front due to the square root singularity. This non-physical property must be somehow reduced by a dissipative process neglected in the Broberg relationship; however, a problem arises even in a non-singular crack model with a constant energy dissipation rate. According to numerical simulations with a friction model equivalent to such dissipation, the peak slip velocity appears an increasing function of r and t [4]. This tendency suggests that the peak slip velocity of $M9$ earthquakes (e.g., duration $T \sim 150$ s for the 2011 Tohoku earthquake) should be 10^4 times greater than that of $M1$ earthquakes (e.g., $T \sim 15$ ms for microearthquakes in a South African gold mine). However, seismic slip inversion analyses revealed that they are both on the same order of 1 m/s (e.g., Ide et al. [17] for $M9$; Yamada et al. [25] for $M1$). Therefore, a model of the dissipation process that restricts slip velocity for such a wide magnitude range is required to understand the physics of earthquakes.

In the following context, we show how dynamic rupture propagation can be modeled in earthquake mechanics. We note the following important properties that have been proposed by observational studies and should be modeled: (1) the rupture velocity is 50–90% of the Rayleigh or shear wave velocity and (2) the slip velocity on faults is finite and scale-invariant. Considering these properties, we focus on friction and off-fault damage to model the dissipation processes previously developed by multiple seismologists. A previous numerical study has already suggested that both friction and damage have a somewhat similar effect on the dissipation rate and self-similarity of rupture propagation [3, 4]. In this study, we analytically illustrate this similar effect by assuming self-similar crack growth.

2 Formulation and Physically Reasonable Models

2.1 Self-similar Displacement, Velocity, and Strain

First, we confirm the following property for a function of $X := x/t$, where $x \in \mathbb{R}^n$ ($n = 2$ or 3) and $t > 0$. Suppose that $\Phi(x, t) = \Phi(X)$ is compactly supported along x-axis, and $\operatorname{supp} \Phi := \{(x, t) \mid \Phi \neq 0\}$ is self-similarly growing inside of $\mathbb{R}^n$. Then,

$$\int_{\mathbb{R}^n} \Phi(X)\mathrm{d}x = t^n \int_{\operatorname{supp} \Phi(X)} \Phi(X)\mathrm{d}X, \tag{6}$$

where t^n comes from Jacobian $|\nabla X|^{-1}$, and the integral with respect to X is independent of t. Via time-derivative, the following also holds:

$$\int_{\mathbb{R}^n} \partial_t \Phi(X)\mathrm{d}x = t^{n-1} \int_{\operatorname{supp} \Phi(X)} \Phi(X)\mathrm{d}X. \tag{7}$$

A complete partial differential equation was given as an initial and boundary value problem for the case of an arbitrary-shaped fault surface embedded in a finite linear elastic body [13]. In this research, we simply consider that an already ruptured region $\Gamma(t) = \operatorname{supp} D$ at time t is a simply connected subset of a planar fault $\left(\subset \mathbb{R}^2\right)$ embedded in an infinite homogeneous elastic and/or inelastic domain $\left(= \mathbb{R}^3\right)$.

Hereafter, we assume a self-similar system that satisfies Eq. (4). The self-similarity can be represented by a homogeneous function of degree N:

$$u(\alpha x, \alpha t) = \alpha^N u(x, t), \tag{8}$$

for an arbitrary positive constant, α, which is satisfied if

$$u(x, t) = t^N u\,(X) \tag{9}$$

holds, where $X := x/t$ again. From the definition of slip, (1),

$$D(x, t) = t^N D(X) \tag{10}$$

also holds, and substituting it into Eqs. (3) and (6) results in

$$M_0(t) = \mu t^N \int_{\Gamma} D(X)dx \propto t^{N+2}.$$

Therefore, through a comparison of this relationship with the empirical law (4), we conclude that $N = 1$.

The spatial- and time-derivatives of Eq. (8) yield that strain $\varepsilon := \frac{1}{2}\left(\nabla u + (\nabla u)^T\right)$ and velocity $v := \partial_t u$ are homogeneous functions of degree $N - 1 = 0$. Hence,

$$v(x, t) = v(X), \tag{11}$$

$$\varepsilon(x, t) = \varepsilon(X) \tag{12}$$

are obtained. Specifically, Eq. (11) indicates the scale invariance of peak slip velocity mentioned in the Introduction if v is finite because of some dissipative process.

2.2 Linear PDE and Reasonable Modeling of Friction

Here, we consider the requirement to reproduce self-similar and scale-invariant peak slip velocity. To avoid the emergence of infinite stress, strain, and velocity in the vicinity of a dynamically propagating rupture front in numerical models, various types of on-fault frictional property have been proposed. Particularly for the co-seismic slip velocity range ($\sim$ 1 m/s), a slip-weakening law

$$f(x, t) = (f_s - f_d)\phi\left(\frac{|D(x, t)|}{D_c}\right) + f_d \tag{13}$$

has been widely employed, where f_s and f_d are the static and dynamic friction, respectively, D_c is the characteristic slip distance, $\phi(s) := (1 - s)H(1 - s)$, and $H(\cdot)$ is the Heaviside function (e.g., [2, 16]). This model is approximately consistent with results from laboratory stick–slip experiments of rock samples above $\sim$ 1 cm/s (e.g., [21]). If Eq. (10) holds with $N = 1$, slip-dependent friction is represented as

$$f(x, t) = f(tD(X)). \tag{14}$$

Also, the fracture energy (i.e., energy dissipation during rupture growth of unit area) is constant under the model (13) because the energy is approximated as $\frac{1}{2}(f_s - f_d)D_c$ [22]. In the following, we show that dynamic crack growth under the slip-weakening friction law will not be self-similar as in Eqs. (8) and (11).

We consider linear elasticity, where the stress change σ and strain are linearly connected via the elasticity tensor C as follows:

$$\sigma(x, t) = \sigma(X) = C\varepsilon(X).$$

We should assume that v and σ are zero for $t \leq 0$ because velocity and stress perturbations are only caused by an earthquake occurring at $t > 0$. Therefore, the governing equation is the following boundary value problem with respect to X:

$$
\begin{cases}
\rho \partial_t v(X) = \nabla \cdot \sigma(X), & X \in \mathbb{R}^3 \backslash \Gamma \\
\partial_t \sigma(X) = C \partial_t \varepsilon(X), & X \in \mathbb{R}^3 \backslash \Gamma \\
\sigma(X) \nu^\Gamma = f, & X \in \Gamma \\
v \to 0, \sigma \to 0, & |X| \to \infty \text{ (i.e., } |x| \to \infty \text{ or } t \downarrow 0),
\end{cases}
\tag{15}
$$

where ρ is the material density and $\sigma \nu^\Gamma$ is the traction change on Γ due to the faulting process. In our model, the fault is a planar shear crack (i.e., $D \cdot \nu^\Gamma = 0$ and ν^Γ is constant), which does not change the normal stress on Γ. Therefore, $\sigma \nu^\Gamma$ is parallel to Γ (i.e., $\left(\sigma \nu^\Gamma\right) \cdot \nu^\Gamma = 0$) and balances with the decrease in friction on the sliding surface f.

So that Eq. (15) represents a self-similar system, $f = f(X)$ must hold, which tells us that any slip-dependent friction (14) cannot completely reproduce self-similar crack growth. Even if Eq. (13) is introduced, f becomes almost constant for a sufficiently larger value of t because of Eq. (14). This means that the problem asymptotes toward the singular crack problem, in which slip velocity diverges. In the same way, the time-weakening friction, $f \propto \phi((t - t_x)/t_c)$, where t_x is the time at which point $x \in \Gamma$ is ruptured and t_c is a characteristic time [3], is also inappropriate for our purpose. One possibility is a velocity-dependent friction $f(V(X))$. The above analysis shows that, for example, velocity-weakening [8] and velocity-strengthening [3] friction models are reasonable for reproducing the scale-invariant peak slip velocity.

2.3 *Energy Conservation Law with a Non-linear Damage Model*

Many geological investigations have reported highly damaged rock surrounding faults with observations of an enormous number of microcracks in fault outcrops (e.g., [7, 10]). The microcrack density per unit area of a cross section decays exponentially as the distance from principal slip plane increases [10]. Therefore, intense co-seismic stress concentration around the dynamic rupture front could generate these microcracks, which could dissipate a non-negligible amount of energy by creating new micro-surfaces. Theoretical modeling consisting of linear elasticity and on-fault friction, as discussed in the previous subsection, has played a crucial role in understanding fault dynamics. However, modeling the damage generation is necessary if it dominates during an actual energy dissipation process.

Several theoretical and numerical models of dynamic damage generation have been proposed. As a realistic model, Yamashita [26] considered each opening microcrack distributed around the fault and executed a finite difference calculation. On the other hand, continuum mechanics-based modeling has been developed by many authors because such discrete modeling is not easy to handle. In the following, we consider a constitutive law including plasticity, which is described as follows:

$$
\partial_t \sigma = C \left(\partial_t \varepsilon - \partial_t \varepsilon^p \right), \tag{16}
$$

where the plastic strain ε^p is zero in the early stage of deformation (i.e., small value of σ) and its rate $\partial_t \varepsilon^p$ depends non-linearly on σ; see Dunham [9] for a specific case of the evolution law of ε^p. By introducing elastic and plastic parts of stress σ as $\sigma^e := C\varepsilon$ and $\sigma^p := C\varepsilon^p$, Eq. (16) can be written as

$$\partial_t \sigma = \partial_t \left(\sigma^e - \sigma^p\right). \tag{17}$$

Laboratory triaxial compression experiments using rock samples have shown that the stress accumulation rate is almost constant during the early stage of loading and decreases in a highly stressed state, as with Eq. (17). Immediately after the experiment, CT imaging revealed an enormous amount of microcracks inside the sample [19]. Therefore, Eq. (17) applies to damage modeling in the highly compressional state.

Here, we discuss a macroscopic energy conservation law including off-fault plastic deformation and on-fault friction. By taking an inner product of velocity, v, and the equation of motion, $\rho\partial_t v = \nabla \cdot \sigma$, we get

$$\frac{\rho}{2}\int\limits_{\mathbb{R}^3\setminus\Gamma} \partial_t |v|^2 \mathrm{d}x + \frac{1}{2}\int\limits_{\mathbb{R}^3\setminus\Gamma} \partial_t \operatorname{tr}\left(\sigma^e\varepsilon\right)\mathrm{d}x = \int\limits_{\Gamma} V \cdot f \mathrm{d}x + \int\limits_{\mathbb{R}^3\setminus\Gamma} \operatorname{tr}\left(\sigma^p \partial_t \varepsilon\right)\mathrm{d}x, \tag{18}$$

where the boundedness of v, σ, and ε is assumed on the basis of a physical requirement. See Hirano [14] for a detailed derivation, i.e., Eq. (18) can be obtained by substituting Eq. (17) into Eq. (3.1) of [14] although the original equation was based on linear elasticity. On the left-hand side, the first and second terms represent the rate of bulk kinetic energy and released bulk elastic strain energy, respectively. On the right-hand side, the first term is the frictional work rate, which represents dissipated energy due to friction, and the second term is similar to the virtual work rate due to the plastic part of the stress. Thus, this second term refers to energy dissipation due to plastic strain or damage.

3 Discussion and Conclusion: Macroscopic Equivalence of Friction and Damage

Both the non-linear model with off-fault damage and the linear elastic model with on-fault friction have contributed to our understanding of self-similar dynamic rupture growth. In this section, we confirm that the energies of these two different models are macroscopically equivalent only if $f = f(X)$ via a dimensional analysis. As with above, v, ε, σ^e, σ^p, and V are assumed to be a function of X for scale invariance.

Given Eqs. (6) and (7), each term in Eq. (18) is proportional to t^2 if $f = f(X)$ holds. Otherwise ($f \neq f(X)$), the first term in the right-hand side is not proportional to t^2, which means that friction does not contribute to the energy dissipation for a sufficiently large value of t. Equation (18) also means that the total released potential energy is proportional to t^3 according to time integration, which was observed seismologically as Eq. (4). Hence, we can conclude the following model patterns:

1. With linear elasticity and any self-similar friction, $f(X)$ (e.g., [8]), the energy balance holds even though the second term of the right-hand side of Eq. (18) is absent.
2. With linear elasticity and slip- or time-weakening friction (e.g., [2–4]), self-similarity is achieved not in the strict sense but as an asymptote. Here, the energy dissipation rate is not proportional to t^2, which implies the divergence of the slip velocity as an analogy of the singular crack problem.
3. With the damage model and self-similar friction (e.g., [9]), both show energy dissipation proportional to t^2 and, therefore, contribute to the energy balance.
4. With the damage model and non-self-similar friction (e.g., [4]), only damage becomes a dominant dissipative process to satisfy the energy balance and self-similarity.

In the above, the cited authors executed 2-D numerical simulations using each type of model, while our analytical modeling is valid also for 3-D cases. A key finding is that patterns 1, 3, and 4 will contribute to reproducing the sub-Rayleigh rupture velocity and scale-invariant peak slip velocity. This conclusion means that we can mimic damage rheology by considering an appropriate friction model, as conducted numerically [4]. On the other hand, seismological observations cannot conclude whether damage or friction is more important for fault mechanics because observations are typically conducted at far-field. Hence, seismological and geological models can be independent, where the former can be achieved even under model pattern 1, while the latter requires model patterns 3 or 4.

Acknowledgments This work was supported by JSPS KAKENHI Grant No. 17H02857.

References

1. Aki, K.: 4. Generation and propagation of G waves from the Niigata earthquake of June 16, 1964. Part 2. Estimation of earthquake moment, released energy, and stress-strain drop from the G wave spectrum. Bulletin of Earthquake Research Institute. University of Tokyo (1966)
2. Andrews, D.J.: Rupture velocity of plane strain shear cracks. J. Geophys. Res. **81**(32), 5679–5687 (1976)
3. Andrews, D.J.: Rupture models with dynamically determined breakdown displacement. Bull. Seismol. Soc. Am. **94**(3), 769–775 (2004)
4. Andrews, D.J.: Rupture dynamics with energy loss outside the slip zone. J. Geophys. Res. **110**(B1), B01307 (2005)
5. Barenblatt, G.I.: Scaling. Cambridge University Press, Cambridge (2003)
6. Broberg, K.B.: Cracks and Fracture. Academic Press (1999)
7. Chester, F.M., Chester, J.S.: Ultracataclasite structure and friction processes of the Punchbowl fault, San Andreas system California. Tectonophysics **295**(1–2), 199–221 (1998)
8. Cochard, A., Madariaga, R.: Dynamic faulting under rate-dependent friction. Pure Appl. Geophys. PAGEOPH **142**(3–4), 419–445 (1994)

9. Dunham, E.M., Belanger, D., Cong, L., Kozdon, J.E.: Earthquake ruptures with strongly rate-weakening friction and off-fault plasticity, Part 1: planar faults. Bull. Seismol. Soc. Am. **101**(5), 2296–2307 (2011)
10. Faulkner, D.R., Mitchell, T.M., Healy, D., Heap, M.J.: Slip on "weak" faults by the rotation of regional stress in the fracture damage zone. Nature **444**(7121), 922–5 (2006)
11. Geller, R.J.: Scaling relations for earthquake source parameters and magnitudes. Bull. Seismol. Soc. Am. **66**(5), 1501–1523 (1976). Retrieved from
12. Hirano, S., Yagi, Y.: Dependence of seismic and radiated energy on shorter wavelength components. Geophys. J. Int. 1585–1592 (2017)
13. Hirano, S.: Integral representation and its applications in earthquake mechanics: a review, In: van Meurs, P. (eds.) Mathematical Analysis of Continuum Mechanics and Industrial Applications II. CoMFoS et al.: Mathematics for Industry, vol. 30. Springer, Singapore (2016)
14. Hirano, S. (accepted article).: Résumé: Energy estimation of earthquake faulting processes. In: Singularity and Asymptotic Behavior of Solutions For Partial Differential Equations with Conservation Law: RIMS Kôkyûroku Bessatsu
15. Houston, H.: Influence of depth, focal mechanism, and tectonic setting on the shape and duration of earthquake source time functions. J. Geophys. Res.: Solid Earth **106**(B6), 11137–11150 (2001)
16. Ida, Y.: Cohesive force across the tip of a longitudinal-shear crack and Griffith's specific surface energy. J. Geophys. Res. **77**(20), 3796–3805 (1972)
17. Ide, S., Baltay, A., Beroza, G.C.: Shallow dynamic overshoot and energetic deep rupture in the 2011 Mw9.0 Tohoku-Oki earthquake. Science **332**(6036), 1426–1429 (2011)
18. Kanamaori, H.: The radiated energy of the 2004 sumatra-andaman earthquake. Geophys. Monograph Ser. **170**, 59–68 (2006)
19. Kawakata, H., Cho, A., Kiyama, T., Yanagidani, T., Kusunose, K., Shimada, M.: Three-dimensional observations of faulting process in westerly granite under uniaxial and triaxial conditions by X-ray CT Scan. Tectonophysics **313**, 293–305 (1999)
20. Meier, M.A., Heaton, T., Clinton, J.: Evidence for universal earthquake rupture initiation behavior. Geophys. Res. Lett. **43**(15), 7991–7996 (2016)
21. Ohnaka, M., Yamashita, T.: A cohesive zone model for dynamic shear faulting based on experimentally inferred constitutive relation and strong motion source parameters. J. Geophys. Res. **94**(B4), 4089–4104 (1989)
22. Palmer, A.C., Rice, J.R.: The growth of slip surfaces in the progressive failure of over-consolidated clay. Proc. R. Soc. A: Math., Phys. Eng. Sci. **332**(1591), 527–548 (1973)
23. Uchide, T., Ide, S.: Scaling of earthquake rupture growth in the Parkfield area: Self-similar growth and suppression by the finite seismogenic layer. J. Geophys. Res.: Solid Earth **115**(11), 1–15 (2010)
24. Udías, A., Madariaga, R., Buforn, E.: Source Mechanisms of Earthquakes: Theory and Practice. Cambridge University Press, Cambridge (2014)
25. Yamada, T., Mori, J.J., Ide, S., Abercrombie, R.E., Kawakata, H., Nakatani, M., Iio, Y., Ogasawara, H.: Stress drops and radiated seismic energies of microearthquakes in a South African gold mine. J. Geophys. Res.: Solid Earth **112**(3), 1–12 (2007)
26. Yamashita, T.: Generation of microcracks by dynamic shear rupture and its effects on rupture growth and elastic wave radiation. Geophys. J. Int. **143**(2), 395–406 (2000)
27. Yoshimitsu, N., Kawakata, H., Takahashi, N.: Magnitude −7 level earthquakes: a new lower limit of self-similarity in seismic scaling relationships. Geophys. Res. Lett. **41**, 4495–4502 (2014)

Isobe–Kakinuma Model for Water Waves

Tatsuo Iguchi

Abstract We consider the initial value problem to the Isobe–Kakinuma model for water waves. The Isobe–Kakinuma model is the Euler–Lagrange equation for an approximate Lagrangian which is derived from Luke's Lagrangian for water waves by approximating the velocity potential in the Lagrangian. The Isobe–Kakinuma model consists of $(N+1)$ second-order and a first-order partial differential equations, where N is a nonnegative integer, and is classified into a system of nonlinear dispersive equations. Since the hypersurface $t=0$ is characteristic for the Isobe–Kakinuma model, the initial data have to be restricted in an infinite-dimensional manifold for the existence of the solution. Under this necessary condition and a sign condition, the initial value problem turns out to be well-posed locally in time in Sobolev spaces. Then, we present a rigorous justification of the Isobe–Kakinuma model as a higher order shallow water approximation in the strongly nonlinear regime. The error between the solutions of the Isobe–Kakinuma model and of the full water wave problem turns out to be of order $O(\delta^{4N+2})$ in the case of the flat bottom and of order $O(\delta^{4[N/2]+2})$ in the case of a variable bottom, where δ is a nondimensional parameter given by the ratio of the mean depth to the typical wavelength and represents shallowness of the water. Therefore, the Isobe–Kakinuma model is a much higher approximation than the well-known Green–Naghdi equations.

1 Introduction

We consider the motion of a water filled in $(n+1)$-dimensional Euclidean space together with the water surface. Let t be the time, $x=(x_1,\ldots,x_n)$ the horizontal spatial coordinates, and z the vertical spatial coordinate. We assume that the water surface and the bottom are represented as $z=\eta(x,t)$ and $z=-h+b(x)$, respectively, where $\eta=\eta(x,t)$ is the surface elevation, h the mean depth, and $b=b(x)$

T. Iguchi (✉)
Department of Mathematics, Faculty of Science and Technology, Keio University, 3-14-1 Hiyoshi, Kohoku-ku, Yokohama 223-8522, Japan
e-mail: iguchi@math.keio.ac.jp

H. Itou et al. (eds.), *Mathematical Analysis of Continuum Mechanics and Industrial Applications III*, Mathematics for Industry 34,
https://doi.org/10.1007/978-981-15-6062-0_13

represents the bottom topography. As was shown by Luke [10], the water wave problem has a variational structure and a Lagrangian was given in terms of the velocity potential Φ and the surface elevation η. Isobe [5, 6] and Kakinuma [7–9] approximated the velocity potential Φ in Luke's Lagrangian by

$$\Phi^{\mathrm{app}}(x, z, t) = \sum_{i=0}^{N} \Psi_i(z; b)\phi_i(x, t), \tag{1}$$

where $\{\Psi_i\}$ is an appropriate function system in the vertical coordinate z and may depend on the bottom topography b, whereas $\phi_0, \phi_1, \ldots, \phi_N$ are unknown variables. The Euler–Lagrange equation for the approximate Lagrangian in terms of $\phi_0, \phi_1, \ldots, \phi_N, \eta$ is the Isobe–Kakinuma model. Different choices of the function system $\{\Psi_i\}$ yield different Isobe–Kakinuma models, and we have to carefully choose this function system in order to obtain a good model for water waves. Here, we adopt polynomials

$$\Psi_i(z; b) = (z + h - b(x))^{p_i}, \tag{2}$$

where $p_0, p_1, \ldots, p_N$ are nonnegative integers satisfying $0 = p_0 < p_1 < \cdots < p_N$. In a practical application, it would be better to choose these indices as

(H1) $p_i = 2i$ $(i = 0, 1, \ldots, N)$ in the case of the flat bottom $b(x) \equiv 0$
(H2) $p_i = i$ $(i = 0, 1, \ldots, N)$ in the case of a general bottom topography.

This choice of the indices is very important to give a rigorous justification of the Isobe–Kakinuma model as a higher order shallow water approximation, although the well-posedness holds for any choice of these indices.

We analyze the linear dispersion relation of the Isobe–Kakinuma model, which will be compared with that of the basic equations for water waves. It is revealed that the Isobe–Kakinuma model under our choice (H1) or (H2) of the function system would be a higher order shallow water approximation to the water waves. Then, we consider the initial value problem to the Isobe–Kakinuma model. Since the hypersurface $t = 0$ is characteristic for the Isobe–Kakinuma model, the initial data have to be restricted in an infinite-dimensional manifold for the existence of the solution. We present the well-posedness of the problem locally in time in Sobolev spaces under this necessary condition and a sign condition, which corresponds to a generalized Rayleigh–Taylor sign condition for water waves. Moreover, as in the case of the full water wave problem, the Isobe–Kakinuma model has conserved quantities, that is, the mass and the total energy, and moreover, the momentum in the case of the flat bottom.

In order to compare the Isobe–Kakinuma model with the full water wave problem more precisely in the shallow water regime, we rewrite the equations in an appropriate nondimensional form introducing a nondimensional parameter δ, which is a ratio of the mean depth to the typical wavelength. There are many approximate models for water waves in the shallow water regime $\delta \ll 1$. Among them, the most famous model is the shallow water equations, which is also called the Saint-Venant equations. As was shown by Iguchi [2] and Alvarez-Samaniego and Lannes [1], the error between the solutions of the shallow water equations and of the full water wave problem is of order $O(\delta^2)$. The Green–Naghdi equations are known to be a higher order shallow water approximation for water waves, and the error is of order $O(\delta^4)$. The Isobe–Kakinuma model turns out to be a much higher approximation for water wares, and the error is of order $O(\delta^{4N+2})$ in the case of the flat bottom and of order $O(\delta^{4[N/2]+2})$ in the case of a variable bottom.

2 Isobe–Kakinuma Model

Luke's Lagrangian for water waves is given in terms of the velocity potential Φ and the surface elevation η by

$$\mathscr{L}_{\mathrm{Luke}}(\Phi,\eta) = \int_{-h+b(x)}^{\eta(x,t)} \left(\partial_t \Phi(x,z,t) + \frac{1}{2}|\nabla_X \Phi(x,z,t)|^2 + gz \right) \mathrm{d}z, \tag{3}$$

where $\nabla_X = (\nabla, \partial_z) = (\partial_{x_1}, \ldots, \partial_{x_n}, \partial_z)$ and g is the gravitational constant, and the action function is given by

$$\mathscr{J}(\Phi,\eta) = \int_{t_0}^{t_1}\!\!\int_{\Omega} \mathscr{L}_{\mathrm{Luke}}(\Phi,\eta)\mathrm{d}x\mathrm{d}t,$$

where Ω is an appropriate region in $\mathbf{R}^n$. Replacing Φ with the approximate velocity potential Φ^{app} given by (1) and (2) in Luke's Lagrangian (3), we obtain an approximate Lagrangian

$$\begin{aligned}
&\mathscr{L}^{\mathrm{app}}(\phi_0,\phi_1,\ldots,\phi_N,\eta) \\
&= \sum_{i=0}^{N} \frac{1}{p_i+1} H^{p_i+1} \partial_t \phi_i \\
&\quad + \frac{1}{2} \sum_{i,j=0}^{N} \biggl(\frac{1}{p_i+p_j+1} H^{p_i+p_j+1} \nabla\phi_i \cdot \nabla\phi_j - \frac{2p_i}{p_i+p_j} H^{p_i+p_j} \phi_i \nabla b \cdot \nabla \phi_j \\
&\qquad\qquad + \frac{p_i p_j}{p_i+p_j-1} H^{p_i+p_j-1}(1+|\nabla b|^2)\phi_i\phi_j \biggr) \\
&\quad + \frac{1}{2} g(\eta^2 - (-h+b)^2),
\end{aligned}$$

where $H = H(x,t)$ is the depth of the water and is given by

$$H(x,t) = h + \eta(x,t) - b(x).$$

Here and in what follows we use the notational convention $0/0 = 0$. Then, the corresponding Euler–Lagrange equation has the form

$$\begin{cases} H^{p_i}\partial_t\eta + \displaystyle\sum_{j=0}^{N}\Big\{\nabla\cdot\Big(\frac{1}{p_i+p_j+1}H^{p_i+p_j+1}\nabla\phi_j - \frac{p_j}{p_i+p_j}H^{p_i+p_j}\phi_j\nabla b\Big) \\ \quad + \dfrac{p_i}{p_i+p_j}H^{p_i+p_j}\nabla b\cdot\nabla\phi_j - \dfrac{p_ip_j}{p_i+p_j-1}H^{p_i+p_j-1}(1+|\nabla b|^2)\phi_j\Big\} = 0 \\ \qquad\qquad \text{for} \quad i = 0,1,\ldots,N, \\ \displaystyle\sum_{j=0}^{N} H^{p_j}\partial_t\phi_j + g\eta \\ \quad + \dfrac{1}{2}\Big\{\Big|\displaystyle\sum_{j=0}^{N}(H^{p_j}\nabla\phi_j - p_jH^{p_j-1}\phi_j\nabla b)\Big|^2 + \Big(\displaystyle\sum_{j=0}^{N}p_jH^{p_j-1}\phi_j\Big)^2\Big\} = 0. \end{cases} \tag{4}$$

This is the Isobe–Kakinuma model that we are going to consider in this article. This consists of $(N+1)$ evolution equations for just one unknown η and one evolution equation for $(N+1)$ unknowns $\phi_0, \phi_1, \ldots, \phi_N$, so that this is an overdetermined and underdetermined composite system. We consider the initial value problem to this Isobe–Kakinuma model (4) under the initial conditions

$$(\eta, \phi_0, \ldots, \phi_N) = (\eta_{(0)}, \phi_{0(0)}, \ldots, \phi_{N(0)}) \quad \text{at} \quad t = 0. \tag{5}$$

3 Linear Dispersion Relation

We linearize the Isobe–Kakinuma model (4) around a rest state $(\eta, \phi_0, \ldots, \phi_N) = \mathbf{0}$ in the case of the flat bottom. Putting $\boldsymbol{\psi} = (h^{p_0}\phi_0, \ldots, h^{p_N}\phi_N)^{\mathrm{T}}$, the linearized equations can be written in a matrix form as

$$\begin{pmatrix} 0 & h\mathbf{1}^{\mathrm{T}} \\ -h\mathbf{1} & O \end{pmatrix}\partial_t\begin{pmatrix} \eta \\ \boldsymbol{\psi} \end{pmatrix} + \begin{pmatrix} gh & \mathbf{0}^{\mathrm{T}} \\ \mathbf{0} & A(hD) \end{pmatrix}\begin{pmatrix} \eta \\ \boldsymbol{\psi} \end{pmatrix} = \mathbf{0}, \tag{6}$$

where $\mathbf{1} = (1, \ldots, 1)^{\mathrm{T}}$ and $A(hD) = -A_0h^2\Delta + A_1$. Here, the $(N+1)\times(N+1)$ matrices A_0 and A_1 are given by

$$A_0 = \Big(\frac{1}{p_i+p_j+1}\Big)_{0\le i,j\le N}, \qquad A_1 = \Big(\frac{p_ip_j}{p_i+p_j-1}\Big)_{0\le i,j\le N}.$$

Therefore, the linear dispersion relation is given by

$$\det\begin{pmatrix} gh & \sqrt{-1}h\omega\mathbf{1}^{\mathrm{T}} \\ -\sqrt{-1}h\mathbf{1} & A(h\xi) \end{pmatrix} = 0,$$

where $\xi \in \mathbf{R}^n$ is the wave vector, $\omega \in \mathbf{C}$ the angular frequency, and $A(h\xi) = (h|\xi|)^2 A_0 + A_1$. We can expand this determinant as

$$h^2\omega^2 \det \tilde{A}(h\xi) - gh \det A(h\xi) = 0. \tag{7}$$

Here and in what follows, we use the notation

$$\tilde{A} = \begin{pmatrix} 0 & \mathbf{1}^{\mathrm{T}} \\ -\mathbf{1} & A \end{pmatrix}$$

for a matrix A. Concerning the determinants in the dispersion relation (7), we have the following proposition.

Proposition 1 *1. For any $\xi \in \mathbf{R}^n \setminus \{0\}$, the symmetric matrix $A(h\xi)$ is positive.*
2. *There exists $c_0 > 0$ such that for any $\xi \in \mathbf{R}^n$ we have $\det \tilde{A}(h\xi) \geq c_0$.*
3. *$(h|\xi|)^{-2} \det A(h\xi)$ is a polynomial in $(h|\xi|)^2$ of degree N and the coefficient of $(h|\xi|)^{2N}$ is $\det A_0$.*
4. *$\det \tilde{A}(h\xi)$ is a polynomial in $(h|\xi|)^2$ of degree N and the coefficient of $(h|\xi|)^{2N}$ is $\det \tilde{A}_0$.*

Therefore, we can define the phase speed $c_{IK}(\xi)$ of the plane wave solution to (6) related to the wave vector $\xi \in \mathbf{R}^n$ by

$$c_{IK}(\xi) = \pm\sqrt{gh\frac{(h|\xi|)^{-2}\det A(h\xi)}{\det \tilde{A}(h\xi)}}.$$

On the other hand, the phase speed $c_{WW}(\xi)$ of the plane wave solution to a linearized equations for the full water wave problem is given by

$$c_{WW}(\xi) = \pm\sqrt{gh\frac{\tanh(h|\xi|)}{h|\xi|}}.$$

Let us compare these two phase speeds. The following theorems were given by R. Nemoto and T. Iguchi [12].

Theorem 1 *If we choose $p_i = 2i$ $(i = 0, 1, \ldots, N)$, then $(c_{IK}(\xi))^2$ becomes the $[2N/2N]$ Padé approximant of $(c_{WW}(\xi))^2$. More precisely, there exists a positive constant C depending only on N such that for any $\xi \in \mathbf{R}^n$ and any $h, g > 0$ we have*

$$\left|\left(\frac{c_{WW}(\xi)}{\sqrt{gh}}\right)^2 - \left(\frac{c_{IK}(\xi)}{\sqrt{gh}}\right)^2\right| \leq C(h|\xi|)^{4N+2}.$$

Theorem 2 *If we choose $p_i = i$ $(i = 0, 1, \dots, N)$, then for any $\xi \in \mathbf{R}^n$ and any $h, g > 0$ we have*

$$\left| \left(\frac{c_{WW}(\xi)}{\sqrt{gh}} \right)^2 - \left(\frac{c_{IK}(\xi)}{\sqrt{gh}} \right)^2 \right| \le C(h|\xi|)^{4[N/2]+2},$$

where C is a positive constant depending only on N and $[N/2]$ is the integer part of $N/2$.

These theorems imply that the Isobe–Kakinuma model under our choice (H1) or (H2) would be a higher order shallow water approximation for the water waves at least in the linear level.

4 Well-Posedness of the Initial Value Problem

We proceed to consider the initial value problem to the Isobe–Kakinuma model (4) and (5). Here, we remark that the Isobe–Kakinuma model has a drawback, that is, the hypersurface $t = 0$ is characteristic for the model, so that the problem cannot be solved in general. In fact, if the problem has a solution $(\eta, \phi_0, \dots, \phi_N)$, then by eliminating the time derivative $\partial_t \eta$ from the equations we see that the solution has to satisfy the relation

$$\begin{aligned} &H^{p_i} \sum_{j=0}^{N} \nabla \cdot \left(\frac{1}{p_j + 1} H^{p_j+1} \nabla \phi_j - \frac{p_j}{p_j} H^{p_j} \phi_j \nabla b \right) \\ &= \sum_{j=0}^{N} \biggl\{ \nabla \cdot \left(\frac{1}{p_i + p_j + 1} H^{p_i+p_j+1} \nabla \phi_j - \frac{p_j}{p_i + p_j} H^{p_i+p_j} \phi_j \nabla b \right) \\ &\qquad + \frac{p_i}{p_i + p_j} H^{p_i+p_j} \nabla b \cdot \nabla \phi_j - \frac{p_i p_j}{p_i + p_j - 1} H^{p_i+p_j-1} (1 + |\nabla b|^2) \phi_j \biggr\} \end{aligned} \tag{8}$$

for $i = 1, \dots, N$. Therefore, as a necessary condition the initial data and the bottom topography have to satisfy these relations for the existence of the solution.

It is well known that the well-posedness of the initial value problem to the full water wave problem may be broken unless a generalized Rayleigh–Taylor sign condition $-\frac{\partial P}{\partial N} \ge c_0 > 0$ on the water surface is satisfied, where P is the pressure and N is the unit outward normal on the water surface. This sign condition is equivalent to $-\partial_z P \ge c_0 > 0$ because the pressure P is equal to the constant atmospheric pressure P_0 on the water surface. By using Bernoulli's law

$$\partial_t \Phi + \frac{1}{2} |\nabla_X \Phi|^2 + \frac{1}{\rho}(P - P_0) + gz \equiv 0,$$

the sign condition can be written in terms of our unknowns $(\eta, \phi_0, \dots, \phi_N)$ and b as $a(x,t) \ge c_0 > 0$, where

$$
\begin{aligned}
a &= g + \sum_{i=0}^{N} p_i H^{p_i - 1} \partial_t \phi_i \qquad (9)\\
&\quad + \frac{1}{2} \sum_{i,j=0}^{N} \Big\{ (p_i + p_j) H^{p_i + p_j - 1} \nabla \phi_i \cdot \nabla \phi_j - 2p_i(p_i + p_j - 1) H^{p_i + p_j - 2} \phi_i \nabla b \cdot \nabla \phi_j \\
&\qquad + p_i p_j (p_i + p_j - 2) H^{p_i + p_j - 3} (1 + |\nabla b|^2) \phi_i \phi_j \Big\}.
\end{aligned}
$$

In fact, we have $-\frac{1}{\rho}\partial_z P^{\rm app} = g + \partial_z \partial_t \Phi^{\rm app} + \nabla_X \partial_z \Phi^{\rm app} \cdot \nabla_X \Phi^{\rm app} = a$ on $z = \eta(x,t)$. The following theorem was given by Murakami and Iguchi [11] in the case $N = 1$ and by Nemoto and Iguchi [12] in the general case and guarantees the well-posedness of the initial value problem to the Isobe–Kakinuma model under the necessary conditions (8) and the sign condition.

Theorem 3 *Let g, h, c_0, M_0 be positive constants and m an integer such that $m > n/2 + 1$. There exists a time $T > 0$ such that if the initial data $(\eta_{(0)}, \phi_{0(0)}, \dots, \phi_{N(0)})$ and b satisfy the relations in* (8) *and*

$$
\begin{cases}
\|\eta_{(0)}\|_m + \|\nabla \phi_{0(0)}\|_m + \|(\phi_{1(0)}, \dots, \phi_{N(0)})\|_{m+1} + \|b\|_{W^{m+2,\infty}} \le M_0, \\
h + \eta_{(0)}(x) - b(x) \ge c_0, \quad a(x,0) \ge c_0 \quad \text{for } \ x \in \mathbf{R}^n,
\end{cases}
$$

then the initial value problem (4) *and* (5) *has a unique solution $(\eta, \phi_0, \dots, \phi_N)$ satisfying*

$$
\eta, \nabla \phi_0 \in C([0,T]; H^m), \quad \phi_1, \dots, \phi_N \in C([0,T]; H^{m+1}).
$$

Remark 1 (1) In the sign condition $a(x,0) \ge c_0 > 0$, we use the quantities $\partial_t \phi_j(x,0)$ for $j = 1, \dots, N$, which should be written in terms of the initial data. Although the hypersurface $t = 0$ is characteristic for the Isobe–Kakinuma model (4), we can express $\partial_t \phi_j(x,0)$ in terms of the initial data and b.

(2) Let us introduce a canonical variable ϕ by

$$
\phi = \Phi^{\rm app}|_{z=\eta} = \sum_{i=0}^{N} H^{p_i} \phi_i.
$$

As we will see later, once we are given the initial data for the canonical variables (η, ϕ), the initial data $\phi_{0(0)}, \dots, \phi_{N(0)}$ for the Isobe–Kakinuma model are uniquely determined so that the necessary conditions (8) are satisfied.

5 Conserved Quantities

As in the case of the full water wave problem, the Isobe–Kakinuma model (4) has conserved quantities: mass and energy, which are given by

$$\text{Mass} = \int_{\mathbf{R}^n} \eta \, \mathrm{d}x,$$

$$\begin{aligned}
\text{Energy} &= \int_{\mathbf{R}^n} \Bigl(\int_{-h+b}^{\eta} \frac{1}{2}\rho |\nabla_X \Phi^{\mathrm{app}}|^2 \mathrm{d}z \Bigr) \mathrm{d}x + \int_{\mathbf{R}^n} \Bigl(\int_0^{\eta} \rho g z \, \mathrm{d}z \Bigr) \mathrm{d}x \\
&= \frac{\rho}{2} \int_{\mathbf{R}^n} \Bigl\{ \sum_{i,j=0}^{N} \Bigl(\frac{1}{p_i + p_j + 1} H^{p_i+p_j+1} \nabla\phi_i \cdot \nabla\phi_j - \frac{2p_i}{p_i + p_j} H^{p_i+p_j} \phi_i \nabla b \cdot \nabla\phi_j \\
&\quad + \frac{p_i p_j}{p_i + p_j - 1} H^{p_i+p_j-1} (1 + |\nabla b|^2) \phi_i \phi_j \Bigr) + g\eta^2 \Bigr\} \mathrm{d}x.
\end{aligned}$$

Moreover, if the bottom is flat, then we have another conserved quantity, that is, the horizontal components of the momentum, which is given by

$$\text{Momentum} = \int_{\mathbf{R}^n} \Bigl(\int_{-h}^{\eta} \rho \nabla \Phi^{\mathrm{app}} \mathrm{d}z \Bigr) \mathrm{d}x = \int_{\mathbf{R}^n} \eta \nabla\phi \, \mathrm{d}x.$$

6 Nondimensionalization

Let us rewrite the Isobe–Kakinuma model (4) in a nondimensional form. Let λ be the typical wavelength and introduce a nondimensional parameter δ by the aspect ratio $\delta = h/\lambda$, which measures the shallowness of the water. We rescale the independent and the dependent variables by

$$x = \lambda \tilde{x}, \quad z = h\tilde{z}, \quad t = \frac{\lambda}{\sqrt{gh}} \tilde{t}, \quad \eta = h\tilde{\eta}, \quad b = h\tilde{b}, \quad \phi_i = \frac{\lambda\sqrt{gh}}{h^{p_i}} \tilde{\phi}_i.$$

Plugging these into (4) and dropping the tilde sign in the notation, we obtain the Isobe–Kakinuma model in the nondimensional form

$$\begin{cases} H^{p_i}\partial_t\eta + \displaystyle\sum_{j=0}^{N}\biggl\{\nabla\cdot\biggl(\frac{1}{p_i+p_j+1}H^{p_i+p_j+1}\nabla\phi_j - \frac{p_j}{p_i+p_j}H^{p_i+p_j}\phi_j\nabla b\biggr) \\ \quad + \dfrac{p_i}{p_i+p_j}H^{p_i+p_j}\nabla b\cdot\nabla\phi_j - \dfrac{p_ip_j}{p_i+p_j-1}H^{p_i+p_j-1}(\delta^{-2}+|\nabla b|^2)\phi_j\biggr\} = 0 \\ \qquad\qquad\qquad\qquad\qquad\qquad \text{for} \quad i=0,1,\ldots,N, \\ \displaystyle\sum_{j=0}^{N} H^{p_j}\partial_t\phi_j + \eta \\ \quad + \dfrac{1}{2}\biggl\{\biggl|\displaystyle\sum_{j=0}^{N}(H^{p_j}\nabla\phi_j - p_jH^{p_j-1}\phi_j\nabla b)\biggr|^2 + \delta^{-2}\biggl(\displaystyle\sum_{j=0}^{N}p_jH^{p_j-1}\phi_j\biggr)^2\biggr\} = 0, \end{cases} \tag{10}$$

where $H(x,t) = 1 + \eta(x,t) - b(x)$ is a normalized depth of the water. As before, we consider the initial value problem to this Isobe–Kakinuma model under the initial conditions

$$(\eta, \phi_0, \ldots, \phi_N) = (\eta_{(0)}, \phi_{0(0)}, \ldots, \phi_{N(0)}) \quad \text{at} \quad t = 0. \tag{11}$$

Then, necessary conditions for the existence of the solution are given by

$$\begin{aligned} &H^{p_i}\sum_{j=0}^{N}\nabla\cdot\biggl(\frac{1}{p_j+1}H^{p_j+1}\nabla\phi_j - \frac{p_j}{p_j}H^{p_j}\phi_j\nabla b\biggr) \\ &= \sum_{j=0}^{N}\biggl\{\nabla\cdot\biggl(\frac{1}{p_i+p_j+1}H^{p_i+p_j+1}\nabla\phi_j - \frac{p_j}{p_i+p_j}H^{p_i+p_j}\phi_j\nabla b\biggr) \\ &\qquad + \frac{p_i}{p_i+p_j}H^{p_i+p_j}\nabla b\cdot\nabla\phi_j - \frac{p_ip_j}{p_i+p_j-1}H^{p_i+p_j-1}(\delta^{-2}+|\nabla b|^2)\phi_j\biggr\} \end{aligned} \tag{12}$$

for $i = 1, \ldots, N$, and the canonical variable ϕ is defined by

$$\phi = \sum_{i=0}^{N} H^{p_i}\phi_i. \tag{13}$$

7 Rigorous Justification of the Isobe–Kakinuma Model

The initial value problem to the full water wave problem in Zakharov–Craig–Sulem formulation in the nondimensional form is written as

$$\begin{cases} \partial_t \eta - \Lambda(\eta, b, \delta)\phi = 0, \\ \partial_t \phi + \eta + \dfrac{1}{2}|\nabla\phi|^2 - \delta^2 \dfrac{(\Lambda(\eta, b, \delta)\phi + \nabla\eta \cdot \nabla\phi)^2}{2(1+\delta^2|\nabla\eta|^2)} = 0, \end{cases} \tag{14}$$

$$(\eta, \phi) = (\eta_{(0)}, \phi_{(0)}) \quad \text{at} \quad t = 0, \tag{15}$$

where $\phi = \phi(x, t)$ is the trace of the velocity potential Φ on the water surface, that is, $\phi(x, t) = \Phi(x, \eta(x, t), t)$ and $\Lambda(\eta, b, \delta)$ is the Dirichlet-to-Neumann map for Laplace's equation. More precisely, the linear operator $\Lambda(\eta, b, \delta)$ depending nonlinearly on the surface elevation η, the bottom topography b, and the parameter δ is defined by

$$\Lambda(\eta, b, \delta)\phi = (\delta^{-2}\partial_z\Phi - \nabla\eta \cdot \nabla\Phi)|_{z=\eta(x,t)},$$

where Φ is a unique solution to the boundary value problem for Laplace's equation

$$\begin{cases} \Delta\Phi + \delta^{-2}\partial_z^2\Phi = 0 & \text{in} \quad -1 + b(x) < z < \eta(x, t), \\ \Phi = \phi & \text{on} \quad z = \eta(x, t), \\ \delta^{-2}\partial_z\Phi - \nabla b \cdot \nabla\Phi = 0 & \text{on} \quad z = -1 + b(x). \end{cases}$$

The following proposition ensures that the initial data $(\phi_{0(0)}, \ldots, \phi_{N(0)})$ for the Isobe–Kakinuma model can be prepared naturally from the initial data for the canonical variables (η, ϕ) in order that they satisfy the necessary conditions (12).

Proposition 2 *Let c_0 be a positive constant and m an integer such that $m > n/2 + 1$. If the initial data $(\eta_{(0)}, \phi_{(0)})$ and the bottom topography b satisfy*

$$\begin{cases} \eta_{(0)} \in H^m, \quad \nabla\phi_{(0)} \in H^{m-1}, \quad b \in W^{m,\infty}, \\ 1 + \eta_{(0)}(x) - b(x) \geq c_0 \quad \text{for} \quad x \in \mathbf{R}^n, \end{cases}$$

then the necessary condition (12) *and the relation* (13) *determine uniquely the initial data $\phi_{0(0)}, \ldots, \phi_{N(0)}$, which satisfy $\nabla\phi_{0(0)} \in H^{m-1}$ and $\phi_{1(0)}, \ldots, \phi_{N(0)} \in H^m$.*

The following theorem was given by Iguchi [3] in the case $N = 1$ and by Iguchi [4] in the general case and gives a rigorous justification of the Isobe–Kakinuma model as a higher order shallow water approximation.

Theorem 4 *Let c_0, M_0 be positive constants and m an integer such that $m > n/2 + 1$ and suppose that* (H1) *or* (H2) *holds. There exist a time $T > 0$ and a constant $\delta_* > 0$ such that if the initial $(\eta_{(0)}, \phi_{(0)})$ and b satisfy*

$$\begin{cases} \|\eta_{(0)}\|_{m+4N+8} + \|\nabla\phi_{(0)}\|_{m+4N+7} \leq M_0 & \text{in the case (H1)}, \\ \|\eta_{(0)}\|_{m+4[N/2]+8} + \|\nabla\phi_{(0)}\|_{m+4[N/2]+7} + \|b\|_{W^{m+4[N/2]+8,\infty}} \leq M_0 & \text{in the case (H2)}, \\ 1 + \eta_{(0)}(x) - b(x) \geq c_0 \quad \text{for} \quad x \in \mathbf{R}^n, \end{cases}$$

then for any $\delta \in (0, \delta_*]$ the initial value problem (14) *and* (15) *to the full water wave problem has a unique solution $(\eta^{\mathrm{WW}}, \phi^{\mathrm{WW}})$ on the time interval* $[0, T]$. *Moreover, for* *any* $\delta \in (0, \delta_*]$ *the initial value problem* (10) *and* (11) *to the Isobe–Kakinuma model with the initial data* $(\phi_{0(0)}, \dots, \phi_{N(0)})$ *determined by Proposition 2 from* $(\eta_{(0)}, \phi_{(0)})$ *has a unique solution* $(\eta^{\mathrm{IK}}, \phi_0, \dots, \phi_N)$ *on the time interval* $[0, T]$. *If we define* ϕ^{IK} *by* (13)*, then for any* $\delta \in (0, \delta_*]$ *and* $t \in [0, T]$ *we have*

$$\|\eta^{\mathrm{WW}}(t) - \eta^{\mathrm{IK}}(t)\|_{m+2} + \|\phi^{\mathrm{WW}}(t) - \phi^{\mathrm{IK}}(t)\|_{m+2} \lesssim \begin{cases} \delta^{4N+2} & \text{in the case (H1)}, \\ \delta^{4[N/2]+2} & \text{in the case (H2)}. \end{cases}$$

Acknowledgments This work was partially supported by JSPS KAKENHI Grant Number JP17K 18742 and JP17H02856.

References

1. Alvarez-Samaniego, B., Lannes, D.: Large time existence for 3D water-waves and asymptotics. Invent. Math. **171**, 485–541 (2008)
2. Iguchi, T.: A shallow water approximation for water waves. J. Math. Kyoto Univ. **49**, 13–55 (2009)
3. Iguchi, T.: Isobe–Kakinuma model for water waves as a higher order shallow water approximation. J. Differ. Equ. **169**, 935–962 (2018)
4. Iguchi, T.: A mathematical justification of the Isobe–Kakinuma model for water waves with and without bottom topography. J. Math. Fluid Mech. **20**, 1985–2018 (2018)
5. Isobe, M.: A proposal on a nonlinear gentle slope wave equation. Proc. Coast. Eng. Jpn. Soc. Civil Eng. **41**, 1–5 (1994) [Japanese]
6. Isobe, M.: Time-dependent mild-slope equations for random waves. In: Proceedings of 24th International Conference on Coastal Engineering, pp. 285–299, ASCE (1994)
7. Kakinuma, T.: [title in Japanese]. Proc. Coast. Eng. Jpn. Soc. Civil Eng. **47**, 1–5 (2000) [Japanese]
8. Kakinuma, T.: A set of fully nonlinear equations for surface and internal gravity waves. In: Coastal Engineering V: Computer Modelling of Seas and Coastal Regions, pp. 225–234, WIT Press (2001)
9. Kakinuma, T.: A nonlinear numerical model for surface and internal waves shoaling on a permeable beach. In: Coastal engineering VI: Computer Modelling and Experimental Measurements of Seas and Coastal Regions, pp. 227–236, WIT Press (2003)
10. Luke, J.C.: A variational principle for a fluid with a free surface. J. Fluid Mech. **27**, 395–397 (1967)
11. Murakami, Y., Iguchi, T.: Solvability of the initial value problem to a model system for water waves. Kodai Math. J. **38**, 470–491 (2015)
12. Nemoto, R., Iguchi, T.: Solvability of the initial value problem to the Isobe–Kakinuma model for water waves. J. Math. Fluid Mech. **20**, 631–653 (2018)

Tsunami-Height Reduction Using a Very Large Floating Structure

Taro Kakinuma and Naoto Ochi

Abstract The tsunami-height reduction using a very large floating structure (VLFS) is discussed, where water waves, interacting with a floating thin plate, are simulated through a numerical model, based on a variational principle, including the nonlinearity and dispersion of waves. First, the tsunami height decreases owing to the wave disintegration, with a creation of shorter floating-body waves of larger phase velocities. Second, after passing the VLFS, the main wave overlaps the shorter waves in the uncovered area, such that the tsunami height increases. Finally, the tsunami height decreases again, as the main wave leaves the shorter waves behind. The final tsunami-height reduction rate increases, as the length or the flexural rigidity of the VLFS is increased. When the VLFS flexural rigidity is larger, it takes a longer time for the VLFS to obtain the final tsunami-height reduction rate.

1 Introduction

A very large floating structure (VLFS), designed for an offshore airport, a storage facility, a wind/solar power plant, an emergency base, or others, has advantages including mobility, as well as friendliness to environment, for seawater can flow under the structure. The interaction between flexible platforms and a fluid can be observed when ice plates are floating at the sea surface (e.g., [9]). In order to design a flexible VLFS interacting with seawater, various numerical calculation methods have been developed: for instance, the Boussinesq-type equations for surface waves were solved by [10] using a finite difference method, to examine the relationship

T. Kakinuma (✉)
Division of Ocean and Civil Engineering, Graduate School of Science and Engineering, Kagoshima University, 1-21-40 Korimoto, Kagoshima, Kagoshima 890-0065, Japan
e-mail: taro@oce.kagoshima-u.ac.jp

N. Ochi
Civil Engineering Department, Ehime Prefectural Government, 4-4-2, Ichibancho, Matsuyama, Ehime 790-8570, Japan
e-mail: k5006633@kadai.jp

H. Itou et al. (eds.), *Mathematical Analysis of Continuum Mechanics and Industrial Applications III*, Mathematics for Industry 34,
https://doi.org/10.1007/978-981-15-6062-0_14

between the bending moment and the flexural rigidity of an elastic thin plate on a progressing solitary wave; [8] studied the interaction of a thin plate with a solitary wave, by coupling a finite element method and a boundary element method; [1] applied a boundary element method for the interaction between thin-plate oscillation and fluid motion, in the coexistence field of linear waves and a current, and simulated the thin-plate response to a moving weight; [5] also studied the cases concerning an airplane moving on a floating airport, using a time-domain mode-expansion method.

When density stratification is developed under a floating structure, its oscillation may generate internal waves, causing change in both the salinity and temperature of seawater, especially in coastal regions through the propagation, shoaling, and breaking of the internal waves. [11] formulated a vertical two-dimensional problem with the framework of a linear potential theory, for the response of a floating thin plate to the field where surface and internal waves coexisted. [4] examined the surface/internal waves due to a moving load on a VLFS in the vertical two dimensions, considering both the nonlinearity and dispersion of the waves.

Conversely, [8] performed hydraulic experiments for a solitary wave traveling through a floating thin plate, where the wave showed disintegration, as its nonlinearity was strong. This result, where the wave height of the solitary wave decreased because of the generation of floating-body waves, suggests that the wave height of a huge tsunami decreases, owing to its propagation through an offshore VLFS. In the present study, we numerically simulate water waves, interacting with a floating thin plate, to discuss the tsunami-height reduction using a VLFS. The governing equations are based on a variational principle [3], for surface waves with both nonlinearity and dispersion, considering the term for the flexibility of a thin-plate floating at the sea surface. The tsunami propagation in a bay has been numerically simulated, to evaluate the tsunami-height reduction due to a VLFS, for various values of its length, as well as flexural rigidity.

2 Governing Equations

The illustration in Fig. 1 is our schematic for the multi-layer system of inviscid and incompressible fluids, represented as i ($i = 1, 2, \ldots, I$) from top to bottom, respectively. The i-layer, the thickness of which is denoted by $h_i(\boldsymbol{x})$ in still water, is sandwiched between two elastic thin plates, where $\boldsymbol{x}$ is the horizontal coordinate (x, y). None of the fluids mixes even in motion, and the density ρ_i ($\rho_1 < \rho_2 < \cdots < \rho_I$) is spatially uniform and temporally constant in each layer. The profiles of the lower and upper interfaces of the i-layer are expressed by $z = \eta_{i,0}(\boldsymbol{x}, t)$ and $z = \eta_{i,1}(\boldsymbol{x}, t)$, respectively, at which the pressure is defined as $p_{i,0}(\boldsymbol{x}, t)$ and $p_{i,1}(\boldsymbol{x}, t)$, respectively. The thin plate touching the upper interface of the i-layer is called the i-plate, the density and vertical width of which are m_i and δ_i, respectively. If both the density m_i and the flexural rigidity B_i of the i-plate are zero, the plate yields no resistance to fluid motion, such that the two immiscible fluids touch each other

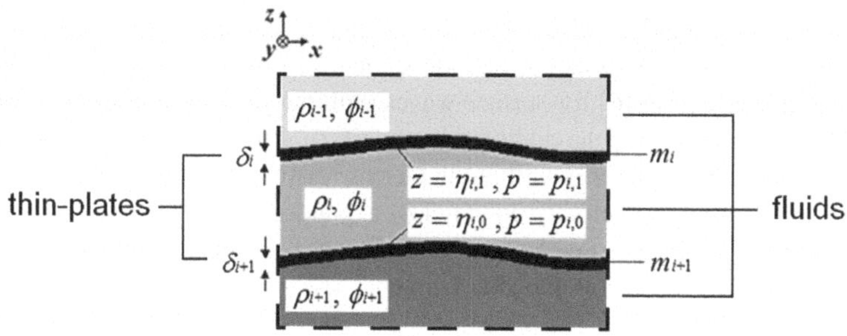

Fig. 1 A schematic for multi-layer fluids sandwiching flexible thin plates

directly without any plate. Surface tension and capillary action are neglected, and the friction is also omitted for simplicity.

Fluid motion is assumed to be irrotational, resulting in the existence of velocity potential ϕ_i, which is expanded into a power series with respect to the vertical coordinate z, with weightings $f_{i,\alpha}$, in a manner similar to that for the derivation of nonlinear surface wave equations by [2], as

$$\phi_i(\boldsymbol{x}, z, t) = \sum_{\alpha=0}^{N_i-1} \left[f_{i,\alpha}(\boldsymbol{x}, t) \cdot z^{\alpha} \right] \equiv f_{i,\alpha} \cdot z^{\alpha}, \tag{1}$$

where N_i is the number of terms for the expanded velocity potential in the i-layer; the sum rule of product is adopted for subscript α; the top face of the 1-layer is described by $z = 0$ in the still water condition.

In the i-layer, if both the displacement of one interface $\eta_{i,1-j}(\boldsymbol{x}, t)$ ($j = 0$ or 1) and the pressure on the other interface $p_{i,j}(\boldsymbol{x}, t)$ are known, then the unknown variables are the velocity potential $\phi_i(\boldsymbol{x}, z, t)$ and the interface displacement $\eta_{i,j}(\boldsymbol{x}, t)$, such that the definition of the functional for the variational problem in the i-layer, $\mathscr{F}_i$, is as follows [3]:

$$\mathscr{F}_i[\phi_i, \eta_{i,j}] = \int_{t_0}^{t_1} \iint_A \mathscr{L}_i[\phi_i, \eta_{i,j}] \, \mathrm{d}A\mathrm{d}t, \tag{2}$$

$$\mathscr{L}_i[\phi_i, \eta_{i,j}] = \int_{\eta_{i,0}}^{\eta_{i,1}} \left[\frac{\partial \phi_i}{\partial t} + \frac{1}{2}(\nabla\phi_i)^2 + \frac{1}{2}\left(\frac{\partial \phi_i}{\partial z}\right)^2 + gz + \frac{p_{i,j} + P_i + W_i}{\rho_i} \right] \mathrm{d}z, \tag{3}$$

where ∇ is a partial differential operator in the horizontal x-y plane, i.e., $\nabla = (\partial/\partial x, \partial/\partial y)$; $(\nabla\phi_i)^2 \equiv |\nabla\phi_i|^2$; the gravitational acceleration g equals 9.8 m/s^2; $P_i = \sum_{k=1}^{i-1}[(\rho_i - \rho_k)gh_k]$ and $W_i = \sum_{k=1}^{i}[(\rho_i - m_k)g\delta_k]$, in case of no buoyancy of structures. The plane A, which is the orthogonal projection of the object domain on to the x-y plane, is assumed to be independent of time. In comparison with the functional referred to in [6] for surface waves with the rotational motion of a fluid, Eqs. (2) and (3) introduce the additional term of $\iiiint (p_{i,j} + P_i + W_i)/\rho_i \, dz \, dA \, dt$ as an interfacial-pressure term, without the terms relating to vorticity.

After substituting the velocity potential ϕ_i expanded as Eq. (1), into Eq. (3), the Euler–Lagrange equations, i.e., Eqs. (4) and (5), on $f_{i,\alpha}$ and $\eta_{i,j}$ are derived for the variational problem of Lagrangian $\mathscr{L}_i$:

$$[\mathscr{L}_i]_{f_{i,\alpha}} \equiv (\mathscr{L}_i)_{f_{i,\alpha}} - \nabla(\mathscr{L}_i)_{\nabla f_{i,\alpha}} - \frac{\partial}{\partial t}(\mathscr{L}_i)_{\frac{\partial f_{i,\alpha}}{\partial t}} = 0, \tag{4}$$

$$[\mathscr{L}_i]_{\eta_{i,j}} \equiv (\mathscr{L}_i)_{\eta_{i,j}} = 0, \tag{5}$$

resulting in Eqs. (6) and (7), respectively, as the fully nonlinear surface/internal wave equations, i.e.,

$$\eta_{i,1}^{\alpha} \frac{\partial \eta_{i,1}}{\partial t} - \eta_{i,0}^{\alpha} \frac{\partial \eta_{i,0}}{\partial t} + \frac{1}{\alpha+\beta+1} \nabla \left[\left(\eta_{i,1}^{\alpha+\beta+1} - \eta_{i,0}^{\alpha+\beta+1} \right) \nabla f_{i,\beta} \right]$$
$$- \frac{\alpha\beta}{\alpha+\beta-1} \left(\eta_{i,1}^{\alpha+\beta-1} - \eta_{i,0}^{\alpha+\beta-1} \right) f_{i,\beta} = 0, \tag{6}$$

where $\alpha = 0, 1, \ldots, N_i - 1$, and the sum rule of product is adopted for subscript $\beta = 0, 1, \cdots, N_i - 1$;

$$\eta_{i,j}^{\beta} \frac{\partial f_{i,\beta}}{\partial t} + \frac{1}{2} \eta_{i,j}^{\beta+\gamma} \nabla f_{i,\beta} \nabla f_{i,\gamma} + \frac{1}{2} \beta\gamma \eta_{i,j}^{\beta+\gamma-2} f_{i,\beta} f_{i,\gamma}$$
$$+ g\eta_{i,j} + \frac{p_{i,j} + P_i + W_i}{\rho_i} = 0 \quad (j = 0 \text{ or } 1), \tag{7}$$

where the sum rule of product is adopted for subscripts $\beta = 0, 1, \ldots, N_i - 1$ and $\gamma = 0, 1, \ldots, N_i - 1$; for instance, $f_{2,3}$ means the weighting of z^3 in the 2-layer. In the derivation process of the equations, no assumption is used for nonlinearity and dispersion of waves, such that the application of this model is expected to be theoretically free from limitations concerning the relative thickness of fluid layers or the frequency band of surface/internal waves.

Conversely, it is assumed that the structures are horizontally very large, i.e., the horizontal length is much larger than the thickness of the thin plates, such that the difference of curvature between the neutral plane, the upper surface, and the lower surface of the thin plates is ignored, resulting in the following classical equation which describes the oscillation of an elastic thin plate:

$$m_i \delta_i \frac{\partial^2 \eta_{i,1}}{\partial t^2} + B_i \nabla^2 \nabla^2 \eta_{i,1} + m_i g \delta_i + p_{i-1,0} - p_{i,1} = 0, \tag{8}$$

where B_i is the flexural rigidity of the i-plate between the $(i-1)$- and i-layers. Even where both the density m_i and the vertical width δ_i are assumed to be constant throughout the i-plate, the flexural rigidity B_i can be distributed along the i-plate.

The physical variables are nondimensionalized using representative wave height H, wavelength λ, water depth h, and fluid density ρ, as

$$\boldsymbol{x}^* = \frac{\boldsymbol{x}}{\lambda},\ z^* = \frac{z}{h},\ t^* = \frac{\sqrt{gh}}{\lambda}t,\ \nabla^* = \lambda\nabla,\ \frac{\partial}{\partial t^*} = \left(\frac{\partial}{\partial t}\right)^* = \frac{\lambda}{\sqrt{gh}}\frac{\partial}{\partial t},$$
$$\eta_{i,e}^* = \frac{\eta_{i,e}}{H},\ m_i^* = \frac{m_i}{\rho},\ \delta_i^* = \frac{\delta_i}{H},\ B_i^* = \frac{B_i}{\rho g \lambda^4},\ p_{i,e}^* = \frac{p_{i,e}}{\rho g h}, \tag{9}$$

where $e = 0$ and 1. The variables in Eq. (9) are substituted into Eq. (8), resulting in

$$\varepsilon^2\sigma^2 m_i^*\delta_i^*\frac{\partial^2\eta_{i,1}^*}{\partial t^{*2}} + \varepsilon B_i^*\nabla^{*2}\nabla^{*2}\eta_{i,1}^* + \varepsilon m_i^*\delta_i^* + p_{i-1,0}^* - p_{i,1}^* = 0, \tag{10}$$

where ε is wave height to water depth ratio H/h, and σ is water depth to wavelength ratio h/λ. In the present paper, it is assumed that the wave nonlinearity is weak in the water of intermediate depth, i.e., the orders of the parameters are $O(\varepsilon) \ll 1$ and $O(\sigma) < 1$; then the first term of the left-hand side of Eq. (10) can be neglected, such that the dimensional equation for the i-plate becomes

$$B_i\nabla^2\nabla^2\eta_{i,1} + m_i g\delta_i + p_{i-1,0} - p_{i,1} = 0. \tag{11}$$

3 Numerical Method

In this and the following sections, the interaction between surface water waves, and a flexible platform floating at the sea surface, is discussed, such that the velocity potential for one layer is described as $\phi(\boldsymbol{x}, z, t) = f_\alpha z^\alpha$, then the unknown variables are the weighting factors f_α and the surface displacement $\eta(\boldsymbol{x}, t) = \eta_{1,1}(\boldsymbol{x}, t)$. The governing equations, i.e., Eqs. (6), (7), and (11), are transformed to finite difference equations, and solved to study the interaction of surface water waves with a floating thin plate, using the implicit scheme developed by [7], for two-layer problems between two fixed horizontal plates.

The numerical results for the surface displacements are in good agreement with the existing experimental data obtained by [8], as shown in the model validation by [4], where the incident solitary wave showed disintegration owing to a floating thin plate. According to the results, where the generation of a preceding short wave was also successfully simulated, the number of terms for the expanded velocity potential, N, is two for the present simulation, such that the balance between the nonlinearity and dispersion of surface waves is considered, for the model takes into account both the linear vertical distribution of horizontal velocity $\boldsymbol{u}$ and the uniform vertical distribution of vertical velocity w.

4 Calculation Conditions

Figure 2 shows the target area for a bay with a VLFS at the water surface, where the still water depth h uniformly equals 50.0 m. The area where 4.0 km $\leq x \leq$ 4.0 km + L and 0.5 km $\leq y \leq$ 3.5 km is covered with the floating thin plate, and the length of the thin plate, L, is 2.0 km or 4.0 km. The flexural rigidity of the thin plate, B, is between 1.0×10^{11} Nm and 1.0×10^{14} Nm, where the flexural rigidity distribution along the x-axis is given as drawn in Fig. 3, to describe the thin-plate covering part of the water area. The lateral boundary condition at $x = 0.0$ km, as well as the shore lines, is the perfect-reflection condition, while the Sommerfeld radiation condition is adopted at $x = 30.0$ km, such that the wave reflection at the bay head is not considered. A vertical wall with the perfect-reflection condition is installed at $y = 2.0$ km, such that half of the target area is the computational domain, where 0.0 km $\leq x \leq$ 30.0 km and 0.0 km $\leq y \leq$ 2.0 km. The grid widths Δx and Δy are 0.05 km and 0.2 km, respectively, while the time-step interval Δt is 4.0×10^{-4} s.

The incident tsunami, the wave height of which is 5.0 m, is the solitary wave solution obtained using the numerical method developed by [12], where the above-described governing equations are reduced for stable water waves, without any thin plate. The initial location of the peak in the surface profile, x_0, is 1.25 km at $t = 0.0$ s, as shown in Fig. 4.

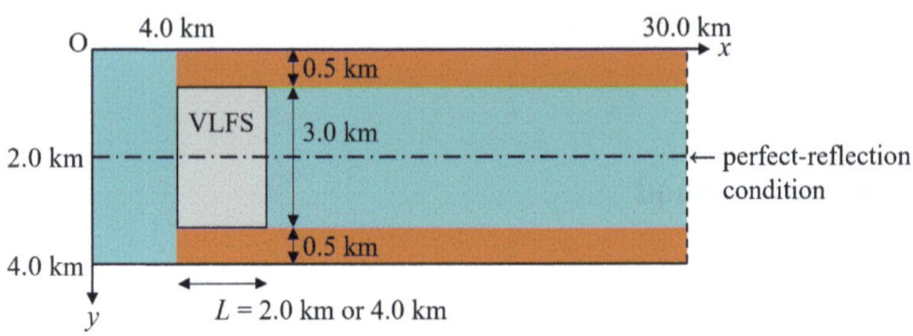

Fig. 2 The target domain for the two-dimensional propagation of a tsunami, interacting with a VLFS, which covers the area where 4.0 km $\leq x \leq$ 4.0 km $+L$, and 0.5 km $\leq y \leq$ 3.5 km; the width of the VLFS is the same as the uniform width of the bay

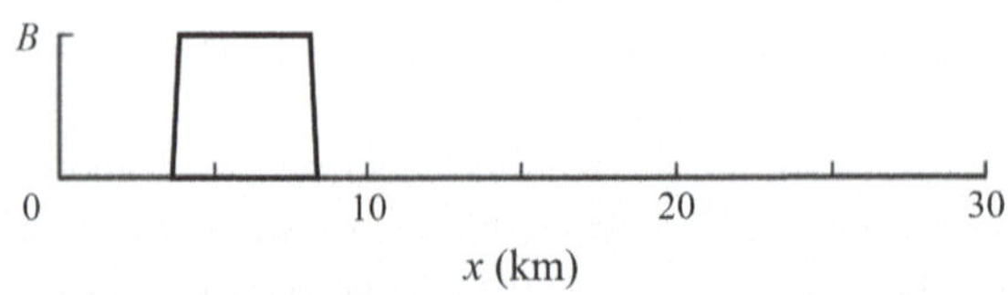

Fig. 3 The distribution of flexural rigidity given along the x-axis, to describe the VLFS covering part of the water area, where the length of the VLFS, L, is 4.0 km

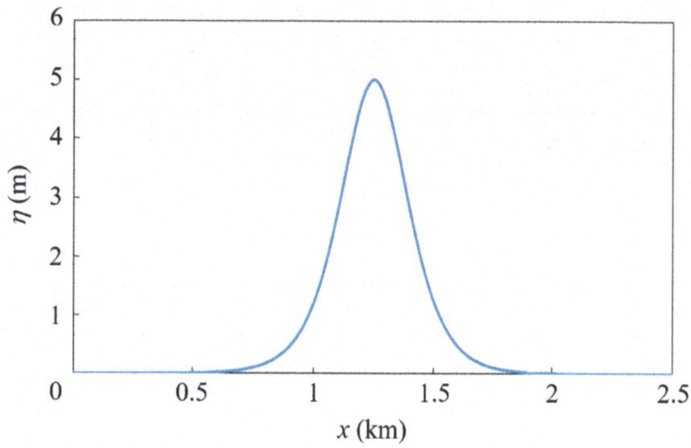

Fig. 4 The water surface profile of the incident tsunami, where η is surface displacement

5 The Effects of the Length and Flexural Rigidity of a VLFS on the Tsunami-Height Reduction

Shown in Fig. 5 is the time variation of the surface profile along $y = 1.0$ km, where η denotes surface displacement; the length L and flexural rigidity B of the VLFS are 4.0 km and 1.0×10^{12} Nm, respectively. Figure 5 indicates that while the solitary wave propagates through the floating thin plate, the wave shows disintegration, generating shorter floating-body waves of larger phase velocities, such that the wave height of the main wave, i.e., the tsunami height, decreases. After passing the thin plate, the main wave overlaps the shorter waves, for the phase velocity of the main wave is larger than that of the shorter waves in the uncovered area, such that the tsunami height increases, especially around $t = 700.0$ s, as shown in Fig. 5. However, as the main wave leaves the shorter waves behind, its wave height decreases, which means that this VLFS is applicable to decreasing the tsunami height.

The time variation of the surface profile along $y = 2.0$ km in the same case as that discussed above is shown in Fig. 6, where the length L and flexural rigidity B of the VLFS are 4.0 km and 1.0×10^{12} Nm, respectively. According to Figs. 5 and 6, the maximum surface displacement is about 5.7 m along both $y = 1.0$ km and 2.0 km, although the maximum wave height of the wave reflected at the bay mouth is larger at $y = 2.0$ km than that at $y = 1.0$ km.

Figure 7 shows the relative maximum surface level $\eta_{1,\max}/\eta_{0,\max}$, for different values of VLFS flexural rigidity B, where the VLFS length L is 4.0 km; $\eta_{1,\max}$ and $\eta_{0,\max}$ denote the maximum surface level at each location, i.e., the tsunami height, with and without a VLFS, respectively. In the cases where the flexural rigidity $B = 1.0 \times 10^{11}$ Nm and 1.0×10^{12} Nm, the final value of the tsunami-height reduction rate $R = (\eta_{0,\max} - \eta_{1,\max})/\eta_{0,\max}$ increases, as the VLFS flexural rigidity B is increased. It should be noted that when B is larger, it takes a longer time for the tsunami-height reduction rate to reach its final value, for the main wave with a

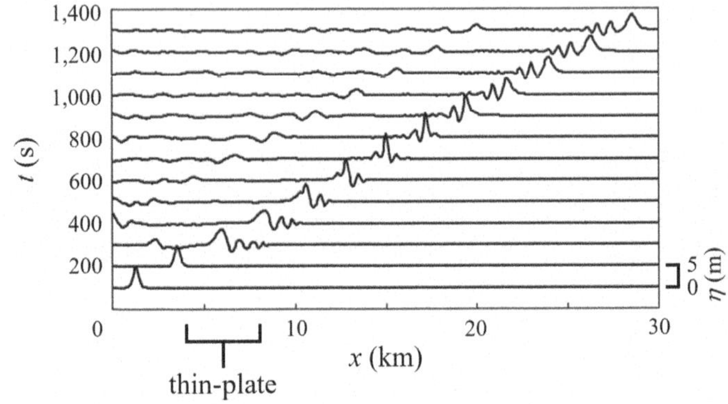

Fig. 5 The time variation of the surface profile along $y = 1.0$ km in the domain shown in Fig. 2, where the length L and flexural rigidity B of the VLFS are 4.0 km and 1.0×10^{12} Nm, respectively; the wave height of the incident wave is 5.0 m, and the still water depth is 50.0 m

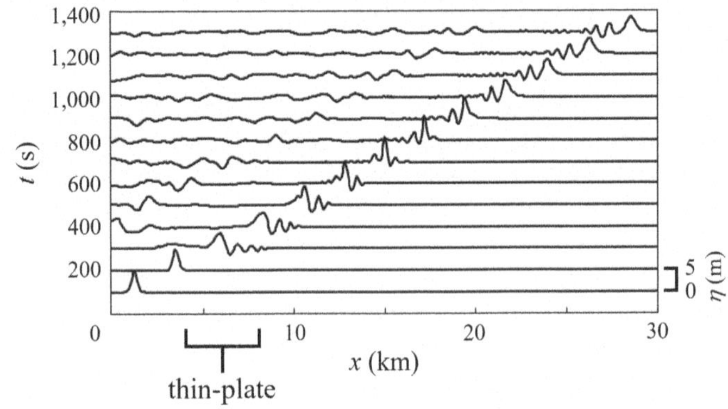

Fig. 6 The time variation of the surface profile along $y = 2.0$ km in the domain shown in Fig. 2, where the length L and flexural rigidity B of the VLFS are 4.0 km and 1.0×10^{12} Nm, respectively; the wave height of the incident wave is 5.0 m, and the still water depth is 50.0 m

slower phase velocity due to a lower wave height should pass a greater number of generated shorter waves.

Depicted in Fig. 8 is the relative maximum surface level $\eta_{1,\max}/\eta_{0,\max}$, for different values of VLFS flexural rigidity B, where the VLFS length L is 2.0 km. The final value of the tsunami-height reduction rate R increases, as the VLFS flexural rigidity B is increased. According to Figs. 7 and 8, in the cases where the flexural rigidity $B = 1.0 \times 10^{11}$ Nm and 1.0×10^{12} Nm, the final value of the tsunami-height reduction rate R increases, as the VLFS is longer.

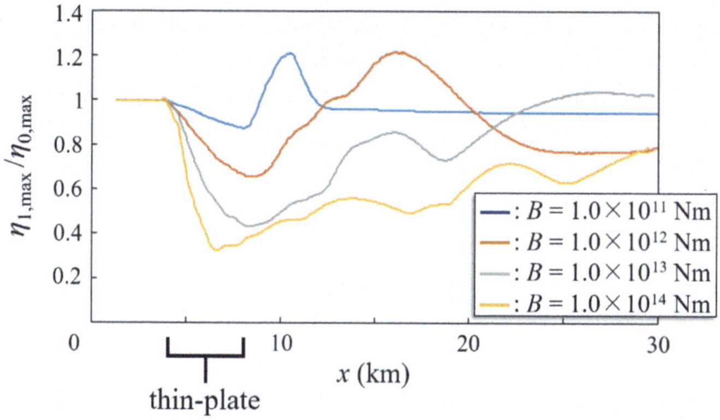

Fig. 7 A relative maximum surface level $\eta_{1,\max}/\eta_{0,\max}$, for different values of VLFS flexural rigidity B, where the VLFS length L is 4.0 km. The initial tsunami height is 5.0 m, and the still water depth is 50.0 m

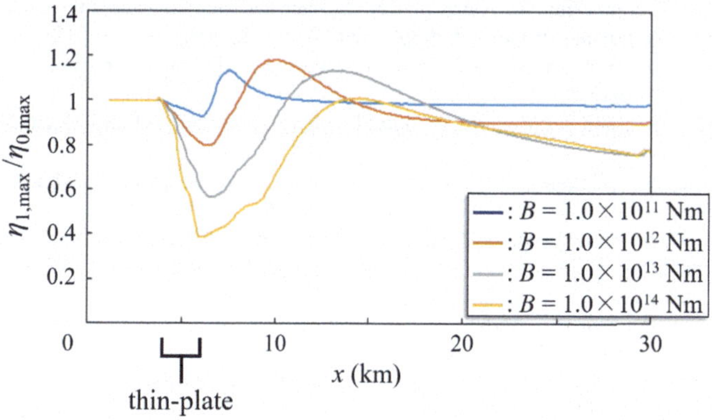

Fig. 8 A relative maximum surface level $\eta_{1,\max}/\eta_{0,\max}$, for different values of VLFS flexural rigidity B, where the VLFS length L is 2.0 km. The initial tsunami height is 5.0 m, and the still water depth is 50.0 m

6 Conclusions

The tsunami-height reduction using a very large floating structure, i.e., VLFS, was discussed, where the water waves, interacting with the floating thin plates, were simulated through the numerical model, based on the variational principle, including the nonlinearity and dispersion of waves.

First, the tsunami height decreased owing to the wave disintegration, with a creation of shorter floating-body waves of larger phase velocities. Second, after passing the VLFS, the main wave overlapped the shorter waves in the uncovered area, such that the tsunami height increased. Finally, the tsunami height decreased again, as the main wave left the shorter waves behind.

The final tsunami-height reduction rate increased, as the length or the flexural rigidity of the VLFS was increased. When the VLFS flexural rigidity is larger, it took a longer time for the VLFS to obtain the final tsunami-height reduction rate, for the main wave with the slower phase velocity due to the lower wave height should pass the greater number of generated shorter waves.

References

1. Hermans, A.J.: A boundary element method for the interaction of free-surface waves with a very large floating flexible platform. J. Fluids Struct. **14**, 943–956 (2000)
2. Isobe, M.: Time-dependent mild-slope equations for random waves. In: Edge, B.L. (ed.) Coastal Engineering 1994, pp. 285–299. ASCE (1994)
3. Kakinuma, T.: A nonlinear numerical model for the interaction of surface and internal waves with very large floating or submerged flexible platforms. In: Chakrabarti, S.K., Brebbia, C.A. (eds.) Fluid Structure Interaction, pp. 177–186. Wessex Insti. Tech. Press (2001)
4. Kakinuma, T., Yamashita, K., Nakayama, K.: Surface and internal waves due to a moving load on a very large floating structure. J. Appl. Math. Article ID 830530, 14 (2012)
5. Kashiwagi, M.: Transient responses of a VLFS during landing and take-off of an airplane. J. Marine Sci. Technol. **9**, 14–23 (2004)
6. Luke, J.C.: A variational principle for a fluid with a free surface. J. Fluid Mech. **27**, 395–397 (1967)
7. Nakayama, K., Kakinuma, T.: Internal waves in a two-layer system using fully nonlinear internal-wave equations. Int. J. Numer. Meth. Fluids. **62**, 574–590 (2010)
8. Sakai, S., Liu, X., Sasamoto, M., Kagesa, T.: Experimental and numerical study on the hydroelastic behavior of VLFS under tsunami. In: Kashiwagi, M., Koterayama, W., Ohkusu, M. (eds.) Hydroelasticity in Marine Technology, pp. 385–391. RIAM (1998)
9. Squire, V.A., Dugan, J.P., Wadhams, P., Rottier, P.J., Liu, A.K.: Of ocean waves and sea ice. Ann. Rev. Fluid Mech. **27**, 115–168 (1995)
10. Takagi, K.: Interaction between solitary wave and floating elastic plate. J. Waterway, Port, Coast. Ocean Eng. **123**, 57–62 (1997)
11. Xu, F., Lu, D.Q.: Wave scattering by a thin elastic plate floating on a two-layer fluid. Int. J. Eng. Sci. **48**, 809–819 (2010)
12. Yamashita, K., Kakinuma, T.: Properties of surface and internal solitary waves. In: Lynett, P.J. (ed.) Coastal Engineering, 2014 Waves. **45**, 15p. ASCE (2015)

Index

H. Itou et al. (eds.), *Mathematical Analysis of Continuum Mechanics and Industrial Applications III*, Mathematics for Industry 34,
https://doi.org/10.1007/978-981-15-6062-0

The manufacturer's authorised representative in the EU is Springer Nature Customer Service Centre GmbH, Europaplatz 3, 69115 Heidelberg, Germany. If you have any concerns regarding our products, please contact ProductSafety@springernature.com

Printed and bound by CPI Group (UK) Ltd, Croydon, CR0 4YY
30/09/2026
02225559-0003